S0-APQ-132

DEAR 60 MINUTES

by Kathleen Fury

A FIRESIDE BOOK
PUBLISHED BY SIMON & SCHUSTER, INC. / NEW YORK

A Fireside Book
Published by Simon & Schuster, Inc.
Simon & Schuster Building
Rockefeller Center
1230 Avenue of the Americas
New York, New York 10020

FIRESIDE and colophon are registered trademarks of
Simon & Schuster, Inc.
Designed by Irving Perkins Associates
Manufactured in the United States of America
10 9 8 7 6 5 4 3 2 1

ISBN: 0-671-50753-2

ACKNOWLEDGMENTS

Many people helped me in compiling this collection. At the offices of "60 Minutes," executive producer Don Hewitt was always helpful and enthusiastic when I managed to interrupt him. So, too, was senior producer Phil Scheffler. As were the "four tigers" and their patient assistants: Mike Wallace, Barbara Dury, Morley Safer, Pam Grant, Ed Bradley, Lisa Deitsch, Harry Reasoner and Jean Dudasik.

Andy Rooney seemed to understand that I felt more comfortable in his cluttered office (so like my own) and tolerated drop-in visits that weren't always strictly necessary. His producer and his editor, Jane Bradford and Bob Forte, were similarly gracious.

Many producers of specific segments provided background information, among them Grace Diekhaus, Steve Glauber, Allan Maraynes, Martin Phillips, Al Wasserman, Bill Willson and Joe Wershba. Palmer Williams, Merri Lieberthal, and Betty Scordamaglia gave me the benefit of their long association with "60 Minutes."

Across the street, I worked closely with Ed Hoppe, executive producer of resources development for CBS News. Ed was my tireless editorial consultant, deadline whip, and cheerleader. Margaret Draper and Sara Maxwell of his staff provided considerable assistance, including directions to the CBS commissary.

Robert Chandler of CBS News, who was vice president

in charge of "60 Minutes" during the period covered by this book, gave the manuscript a careful reading and made a number of good suggestions.

Marjorie Holyoak and her staff at Audience Services explained the complicated, backstage role they play at CBS—and early on warned me that they knew better than anybody, that looking at all that mail would be a major undertaking.

Susan Victor of Simon & Schuster guided the book through to publication with admirable care and professionalism and an astute editorial judgment. I was especially pleased that she agreed not to "clean up" the letters as much as an editor might an ordinary text—for example, to allow writers who use capital letters for emphasis to do so here, or to give the ampersand-lover free rein.

Susan deIpolyi, word processor extraordinaire, "input" the letters with remarkable accuracy, Dwight Stagg helped cull some of them, and Sam Klein of the U.S. Postal Service explained how a letter gets from your house to "60 Minutes."

I would like to thank my cat Rona, fond of chewing paper, for not eating a single one of the thousands of letters that littered the house for months. And Len Fury, for listening, supporting and adding (as he usually does) to my pleasure.

My final and special thanks to Mike Wallace, who suggested that I be considered for this interesting project.

CONTENTS

INTRODUCTION

You will meet people you like in this book; I'm sure of it. Such an incredible diversity of opinion is expressed in the hundreds of letters published here that you're bound to find somebody who feels just as you do about elderly drivers, gun control, lawyers, the Mafia, doctors, American cars—even Pavarotti's voice, barking dogs, and the merits (or lack thereof) of San Francisco.

You will also meet people who remind you of, perhaps, your brother-in-law or the lady next door—those folks whose opinions strike you as demented because they disagree with yours.

But best of all, you'll meet some people who sat down and wrote thoughtful letters on stories covered by "60 Minutes"—letters that shed such new and interesting light you'll reconsider your own feelings. You will also discover profoundly touching personal accounts, as in the agonized letter from a man who believes his baby may have been born deformed because his pregnant wife ate wild raspberries that had been sprayed with weedkiller.

Whether you read this book straight through or dip into sections here and there, you will get a unique glimpse into the soul of America. More than any poll or study could, these letters reveal who we are and what matters to us. We tend to be a self-critical people, so you may be surprised to discover some of the threads that connect the letters: a passionate sense of justice, a desire to see others treated fairly, and a never-ending need to find answers to the question: Just what does the United States of America stand for?

This book is a kind of national town meeting taking up that question as it applies to all kinds of issues, big and small, facing us today. The voices raised here are as diverse as America: young and old and middle aged, rich and poor, rural and urban and suburban. Listened to as a whole, they make a sound like the singing of an enormously energetic, if difficult to conduct, chorus. They make me proud to be an American.

The legendary CBS newsman Edward R. Murrow once told a convention of fellow broadcasters, "I am entirely persuaded that the American public is more reasonable, restrained, and mature than most of the broadcast industry's planners believe. Their fear of controversy is not warranted by the evidence." "60 Minutes" has not feared controversy, which may be one reason for its broad appeal. And Mr. Murrow would probably be confirmed in his high opinion of the American public if he could read the spirited letters written to a news-magazine broadcast he never lived to see.

The letters in this book are a small but representative sample of the almost 150,000 received by "60 Minutes" over the last two years. They have been edited for space, clarity, and grammar (unless corrections would have destroyed something essential in the writer's tone or style). Most of them have never been broadcast or published before; the "letters segment" of "60 Minutes" allows time for only a few remarks from viewers.

In choosing this sampling of the letters, my first criterion was that they make interesting reading. The second was that they be reasonably representative of all the mail on a particular subject. As I worked my way through mounds of material, it soon became clear what types are most interesting to read. Letters that are personal, that tell stories, are generally livelier than those that make abstract points. Fan letters are boring.

I took a cue in this from "60 Minutes" executive producer Don Hewitt, who originated the program in 1968 and remains the creative force behind it. Hewitt himself picks the letters that are read on the broadcast. It's because of his almost unerring instinct for the lively that you never hear a letter read on "60 Minutes" that says, "I love your program and think you do a great job in the service of investigative journalism." They get such letters, but Hewitt finds them self-serving and, worse, dull. (In fact, Hewitt claims to have fired somebody once for using the words "investigative journalism" in a "60 Minutes" promotion.)

I eliminated letters from cranks, but chose lots from viewers who were feeling cranky about particular issues. And I eliminated unsigned letters, making exceptions only when they were compelling enough to warrant inclusion or when the writer had good reason to request anonymity.

I included a letter from somebody who lives in the small town in Indiana where I grew up. That's the sort of thing that makes editing fun.

Andy Rooney once noted, "There are almost no people who are not dentists who can fix teeth, but there are a lot of people who aren't professional writers who write very well." After reading thousands of letters, I agree. Still, like most writing they needed some editing. What I have *not* done is include standard editorial marking—brackets to indicate that a word has been added, ellipses (dots) to indicate words that have been cut. The book would have been virtually unreadable with all those marks. This is a departure from "60 Minutes' " policy, where every change in a letter shown on the screen is indicated. It's an admirable policy, but it drives me nuts trying to figure out what they left out and what they put in. In reading the mail, I discovered I am not alone in this. Here is a letter, printed in its entirety (I wouldn't dare edit it) one viewer wrote:

Dear 60 [sixty] Minutes:

I [see my name below] am confused [don't understand] at why every letter you air [telecast] about previous [sixty minutes] programs has all those brackets [()]. Is it [the brackets] because people [viewers who respond] do not know how to write? Or is it [again the brackets] because you [sixty minutes correspondents] have to read between the lines to make a letter make sense?

Sincerely, [or otherwise]

James [Jim] P. [won't discuss middle name] Hollis
Longwood, FL

P.S. Can I write again about . . . . all those dots?

I agree with Mr. Hollis. And I hope that the letter writers published in this book will find that I have been true to their meaning in my editing.

The process of typesetting destroys the individual look of letters, of course. But if you have a visual imagination, you might bear in mind while reading them how diverse they look: they are handwritten, block printed, typed, and printed on dot-matrix printers. They arrive on lined legal paper, engraved letterheads, notes imprinted "From the desk of . . . ," torn-off pieces of shopping bags, continuous-feed computer paper with perforated edges, pastel stationery illustrated with flowers and bunnies, official letterhead snitched from the office, three-ring binder paper, postcards plain and postcards picture, hotel stationery, "Things to Do Today" notepaper, and much more. Some are perfumed; at least one smelled of mothballs. In these pages only the perfume of their incredible diversity, curiosity, and energy lingers on.

The power of a wafer or a drop of wax or gluten to guard a letter, as it flies over sea, over land, and comes to its address as if a battalion of artillery brought it, I look upon as a fine meter of civilization.
—RALPH WALDO EMERSON

CHAPTER 1

9:00 A.M. on West 57th Street

Every weekday morning at about 9:00 A.M. a red-white-and-blue U.S. postal truck pulls up to the loading dock at 524 West 57th Street in New York. This low-slung, red-brick building, stretching for most of a long city block, houses production and operations for CBS News, local and network. Across the street and catty-corner to it, in a glass mini-skyscraper, are the offices of "60 Minutes."

The streets in this relatively remote part of Manhattan are still quiet at this hour; frequently, the fog that collects overnight on the nearby Hudson River has not yet burned off. The official work day at "60 Minutes" runs from 10:00 to 6:00, but Andy Rooney, a confessed morning person, is probably already in the famous office he once likened to a chicken coop, having coffee with his editor.

Correspondents Mike Wallace, Morley Safer, Ed Bradley, and Harry Reasoner, when not out of town on assignment, are getting ready for the workday. Wallace, an early riser, may still be at home in his East Side brownstone, having coffee and reading the New York Times. Safer may be in the office already; if he has copy to write, he likes to do it during this peaceful time before the phones start ringing. Bradley is probably just finishing his morning workout at a gym. Reasoner, who cherishes a profound hatred for the kind of people who can get up and start *doing* something right away, will arrive around 10:00, having been "up for three hours but active for about twenty minutes."

The brief quiet that precedes the beginning of the workday at one of the world's largest and busiest news centers is broken every day by the sound of metal wheels scraping across pavement. It is the mail being unloaded from the postal truck in huge wheeled tubs filled with canvas sacks full of letters. Roughly half of this mail, an astounding percentage of all the correspondence to CBS News and its many divisions, is addressed to "60 Minutes." In the last two years alone (1982–83), nearly 150,000 people wrote to "60 Minutes," and the letters continue to come in.

The letters "60 Minutes" gets, moreover, reflect the unique nature of the relationship that has grown up between the program and its audience over the remarkable span of time—sixteen years—it has been on the air. Because of its investigative reporting, "60 Minutes" is seen by many Americans as a kind of court of last resort—the place where you turn for help when you've exhausted every other avenue, from newspapers to government officials. Thus, "60 Minutes" may get more mail from prison inmates than their wives and lovers do. "If every prisoner who wrote us saying he was framed were released tomorrow," notes senior producer Phil Scheffler, "the jails would be empty."

But more often it is "ordinary people," those of us whose closest brush with the judicial system is a parking ticket, who write to "60 Minutes." We are a large and devoted audience, keeping the broadcast close to the top of the ratings week after week. In return, we have high expectations. So we write to take "60 Minutes" to task when we think they're wrong, to praise them when we think they've exposed injustice or brought an important issue to national attention. We write to correct their grammar, to comment on their style (even, occasionally, on the clothes the correspondents wear), or to offer our opinion on their journalistic evenhandedness or lack of it. We write to tell them how a segment touched us. Most of all, we write to tell them *our* stories.

We write because the intimacy of the medium makes us feel, somehow, that we know these people. Children have grown from first-graders to first-time parents during the sixteen years Mike Wallace has been on "60 Minutes." He and his colleagues are like family.

As a result, much of the mail the correspondents get is familiarly addressed: "Dear Harry," "Dear Ed," "Dear Andy," "Dear Morley" (though the spelling of dear Morley's name presents a problem to many viewers, who address him in an infinite variety of ways: Maury, Morny, Wally, Murray, Marly, and much, much morely). Mike Wallace, perhaps because of his reputation as one of television's toughest interviewers, seems the one most often addressed with the honorific "Mr."

Executive producer Don Hewitt, who created "60 Minutes," believes the letters constitute "the greatest poll in America." Whether they are statistically representative or not, there's no question they are deeply felt, often funny, insightful, and fascinating to read. Through them, we see ourselves revealed as individuals and as Americans—armed with enthusiasm, spirit, and strong opinion as we head for the twenty-first century.

As long as there are postmen, life will have zest.

—WILLIAM JAMES

CHAPTER 2

One Letter

Though there is really no such thing, imagine a typical letter to "60 Minutes." How does it get there, and what happens to it when it does?

Once written, it must be addressed. A careful observer of the television screen will have noticed and written down the address as it appears on-screen during the letters segment at the end of the broadcast.

Of course, some viewers are more observant than others:

Dear 60 Minutes:

I've noticed, at the end of your program when you read the letters from your mail bag, that the letter you show is an airmail envelope with a "C" stamp on it. Domestic airmail does not exist anymore. The post office does not sell "C" stamps anymore. I feel you should use a regular envelope and one of the more attractive commemorative stamps or a regular 20-cent stamp.

Sincerely,
Scott Stokley (12 yrs. old)
Otis, Oregon

Since Scott—and others—wrote, the illustration has been changed to show a twenty-cent stamp. (Quite possibly by the time this book comes out it will need to be changed again.)

Other viewers, who are suddenly inspired to write, can't find pen and paper fast enough to record the address—and eventually write "60 Minutes" to complain that it doesn't remain on-screen long enough.

Still others care not a whit about such formalities. I have seen envelopes addressed "Mike Wallace, 60 Minutes"—nothing more—that got delivered. The possibilities are endless, and sometimes amusing: "Mickey Rooney, NBC-TV, New York" actually has a good chance of reaching Andy, even though addressed to the wrong network and the wrong Irishman. This is a testament both to the extraordinary fame television brings and to the presence, despite mechanization and zip codes, of real human beings still at work in the Post Office.

If the letter is mailed from more than 250 miles away from West 57th Street, it will travel from its home post office on an airliner to Kennedy airport, then by truck into Manhattan for sorting, and then by an even smaller truck to 524 West 57th Street, where it becomes the responsibility of CBS. According to the Postal Service, if it's mailed before 5:00 P.M. Monday in, say, Cleveland, it will be delivered to the CBS News mailroom by 9:00 A.M. Wednesday morning.

If our imaginary letter is addressed to a specific person by name, it goes to his or her office. Otherwise the receptionist gets it, opens it, and sorts by category. If it's a letter about last Sunday's broadcast and arrives before Thursday afternoon, it has a good chance of being read by the boss, Don Hewitt, who chooses the letters to be read on the air the next Sunday.

A crucial question: Will this be a one-way correspondence, or will the letter be answered? There are several possibilities. If this letter is addressed to Andy Rooney and requests that he send an old sock to a charity auction, he may send it to the wastebasket. Andy Rooney does not see any point in sending out his old socks or responding to people who ask for them.

But if the letter happens to touch a chord with any of the correspondents, it may get a personal response. Each of them has his own method of answering mail; for more information on this, see the chapters entitled "Dear Mike," "Dear Morley," and so on.

The vast majority of the mail, of course, cannot be answered with a personal reply. Responding to these letters is the job of a large department, Audience Services, presided over with friendly efficiency by English-born Marjorie Holyoak. Audience Services is at a different address, across town (CBS is a *large* organization). All day long a small van, affectionately called "the scooter," travels back and forth from West 57th Street to "Black Rock," the corporate headquarters on Avenue of the Americas. Our imaginary letter, if it hasn't been answered at "60 Minutes," will ride the scooter with many others over to Audience Services.

Twenty people work there, six of them assigned solely to "60 Minutes" mail. They read every single letter and decide which of hundreds of possible form replies it should receive, keeping track of how many letters come in and noting whether they are pro or con a particular position.

Audience Services has form-letter replies, with coded numbers, for every conceivable situation—and when an inconceivable situation comes up, they can write a new form letter. There's a form letter for people who saw something on another network but addressed their letter to "60 Minutes" ("We appreciate your taking the time to write to us; however, the broadcast you mentioned was not a CBS presentation"). There's a form letter for people who want a subscription ("It is gratifying to learn of your interest. . . . We would, however, like to correct a misunderstanding. There is no magazine entitled '60 Minutes'—the phrase 'news magazine' simply describes the format").

Though living, breathing people read every single letter, the replies are typed out by printers connected to computers, a concession to progress that pleases those who might otherwise have to do the typing.

Viewers write to "60 Minutes" not only to praise or criticize what they've seen on the air but to suggest what they would *like* to see. In 1983, more than 15,000 people wrote to suggest stories. As Mike Wallace has said, viewers act as "stringers" for the show; it is impossible for the staff to know of every dramatic, newsworthy situation in the country.

Very few of these letters lead to a broadcast, though all are looked at by producers and correspondents. Sometimes a story quite clearly has grown out of a letter; more often, the origins of the lead are murky, because so many sources are available to "60 Minutes." But the letter from a viewer was one of the instrumental factors in the decision to investigate the story of Lenell Geter, a black engineer sentenced to life imprisonment in Texas for the holdup of a Kentucky Fried Chicken restaurant. That broadcast, which earned "60 Minutes" the prestigious Peabody Award, won Lenell Geter his freedom.

The letter was written by a 43-year-old housewife, who signed herself Mrs. Dan Stilwell. She had read about the Geter case in the *Fort Worth Star-Telegram*. In 1982, Lenell Geter was one of six black graduates of South Carolina State College recruited by E-Systems, a major defense contractor, to work in their Greenville, Texas offices. According to legend, the small town once displayed a sign proclaiming itself the home of "The Blackest Soil, the Whitest People."

Geter, who earned $24,000 a year at E-Systems, had been in Greenville only six months when he was arrested in connection with a $615 robbery at a nearby Kentucky Fried Chicken restaurant. Geter, who refused to plead guilty and maintained his innocence throughout, was brought quickly to trial. He was convicted by an all-white jury even though there was no physical evidence connecting him to the crime and co-workers testified he had been at work that day and couldn't possibly have committed it.

With Lenell Geter in jail, friends, colleagues, and other interested parties mounted a campaign for a retrial. Coworkers raised thousands of dollars for his defense. The NAACP got involved. The Dallas newspapers told the story; their accounts led to further coverage around the country. The *New York Times* covered it. The "media guns" went national when the case was discussed on "The Phil Donahue Show" and written up in *People* magazine. But despite all that publicity, Lenell Geter was still behind bars, with his promising life turning to despair, when Mrs. Dan Stilwell rolled a piece of paper into her typewriter and wrote to "60 Minutes."

She is a white woman, a wife and mother who confesses that even during the height of the civil rights movement, she had "never given a lot of thought to the subject of prejudice. I never marched or anything like that. Even writing that letter was out of character for me." But Mrs. Stilwell felt outraged about what had happened to Mr. Geter; she herself had lived in a small Texas town for a brief unhappy time and Greenville reminded her of it. But most of all, Geter's story offended her sense of justice: "It just got my dander up. They have the same rights that we do. It was so unfair—he lost sixteen months of his life that he will never regain."

She thought of "60 Minutes" right away. On the television listings page of her newspaper, she saw a list of network addresses. A former secretary, she composed her letter and typed it neatly, then carefully photocopied the newspaper story, cutting it up and taping it together. She mailed all this to the name in the paper: Gene Jankowski, president of the Columbia Broadcast Group.

"I am not associated with any party of this case," she wrote. "I am just an interested, concerned citizen, and I am irate at the treatment this young man received. Guilty or not, this man's rights were violated. What I would like to see is for '60 Minutes' to investigate this situation. Maybe the national exposure would help Mr. Geter in some way." As she mailed it, she thought, "If only they'll just take the time to read that story. . . ."

Her letter and clipping arrived at CBS corporate headquarters late in February 1983—six months after Geter's arrest and imprisonment. Jankowski passed it along to Van Gordon Sauter, president of CBS News, who sent it to Robert Chandler, then senior vice-president in charge of the "60 Minutes" unit. Chandler read Mrs. Stilwell's letter and the long newspaper piece she enclosed and sent it over to Don Hewitt with a note: "Though complicated, this story could be a pretty good one."

Others had written directly to "60 Minutes" about Lenell Geter already—including his mother and a lawyer for the NAACP. A CBS producer had been collecting a file on the case for possible use. But it was Mrs. Stilwell's letter that, to the best of his recollection, brought the case to Hewitt's attention.

Morley Safer and his crew went to Greenville to look into the story. Their broadcast on the incredible trial of Lenell Geter aired on December 4, 1983. The Dallas Cowboys were playing out of town that day, so most fans were at home watching the game on TV. Those who left their sets tuned to CBS saw Safer unravel a story that did not reflect especially well on the Texas judicial system.

Dallas District Attorney Henry Wade had turned his TV off after the game, and was getting ready to go out when a colleague called him. "'60 Minutes' is doing the Lenell Geter story," he said. Wade turned the set back on and watched it himself.

In the months to follow, Wade would take every possible opportunity to denounce the broadcast, calling it slanted and distorted. But he could not avoid the private reactions of those around him. "That next night," he recalls, "my family was over for dinner. Both my son and my son-in-law wanted to know how we could have convicted Geter. And some people I play golf with asked me the same thing."

His denunciations notwithstanding, ten days after the "60 Minutes" broadcast DA Wade released Lenell Geter from jail and ordered a new trial. By then, Geter had spent sixteen months in prison. Grudgingly, Wade now concedes that "that broadcast had a lot to do with it." One of Geter's lawyers, Ed Sigel, is more outspoken: "'60 Minutes' buckled one of the most powerful men in Texas. What Wade did has never been done before—you just won't find it in the legal history of Texas. If you get over fifteen years in Texas you go straight to the penitentiary. You can't make bond. And you stay in jail until your appeal is reversed, if that ever happens. If it hadn't been for '60 Minutes,' Lenell Geter would still be in jail."

Three months later the case against Geter was dropped altogether. "I feel so fine," says the young engineer, "that if I had feathers I could probably fly. My case got a lot of attention because I'm educated and I have a good job. I just wonder how many others have been arrested and railroaded through like me and we've never heard of them."

Mrs. Stilwell missed the broadcast. She heard about it later that week in the government class she was taking at a local junior college. Perhaps the most important lesson she has learned from all this is that the power of a free press is proportionate to its communication with its audience. And that one letter, like one vote, can indeed make a difference.

CHAPTER 3

Gardening

(*There'll Always Be an America*)

In November 1982, "60 Minutes" first aired a segment called "Plumbago Carpensis and All That." In it, Morley Safer—who is Canadian-born and lived for some years in England and often does stories on English eccentricities—traveled to England to look into the subject of gardening. For fourteen minutes the screen was lit up by the brilliant colors of English gardens in full bloom, by London's annual Chelsea Flower Show, and by scenes of a small police station ablaze with bobby-tended blossoms.

In the segment Safer also visited "Gardener's Question Time" (one of Britian's most popular radio programs), a Lady of the Realm who worried about keeping her ancient garden weed-free, and a gentleman with a laboratory in Cornwall who for fifty years has been zapping his plants with electricity to see if he can make them grow faster.

"Plumbago Carpensis and All That" is a classic "soft" piece: humorous, apolitical, close to daily life, a pleasant change of pace from news of man's inhumanity to man. It appeared between a segment on an American who sold secrets to the Russians and one on French terrorism. ("A Few Minutes with Andy Rooney" wasn't on the broadcast that particular night.)

Those who think that a story about flowers is hardly the

stuff of controversy gravely underestimate the passions of the "60 Minutes" audience.

The piece about British gardening was highly amusing, tops, deft, droll as can be—in short, just swell. There's no doubt now that you've just about cornered the market for the light and memorable American video news feature.

Happy holidays, and cheerio!

Bill Swing
News Director
KPTV 12
Portland, Oregon

Gardening in England will become a milestone in the history of investigative journalism. The segment had it all—passion, excitement, controversy, poignancy! The companion pieces on Terrorism in France and Espionage in America were dwarfed by comparison.

May I be so bold as to suggest some equally worthwhile stories you might pursue in future:

1. A short history of lint.
2. The black market in grocery coupons.
3. What ever happened to Miss Arkansas 1964?
4. A profile of Bert Convy, etc., etc. (Yawn.)

Michael Rende
Toronto, Ontario

Could you please cut down on Morley's bloody Anglophilia! From back-gardens, croquet-fields, and bleedin' royal photographers and other like subjects, he is building up a large migraine headache in the viewing public! Now mind you, I am not averse to Britain, merely getting averse to Morley's sycophantine dotage! So when next Morley feels "the urge" to be quaint and anachronistic then please send him off to Dubrovnik or Dubuque or somewhere, anywhere, but not Britain!

Egon Eigensinn
Los Angeles, California

Ah, to have a garden, now that England's here.

A few years ago, with permission from my super, I planted some seeds in the front yard and less than a month later he'd pulled them all out, saying they were nothing but weeds.

I for one ate my heart out with envy of all those gardeners with their lovely pieces of earth.

Joan Beth Brown
Brooklyn, New York

My heart *sang* when your British guest called organic gardeners cranks! It's about time someone said that in public.

Jo M. Peshek
Slidell, Louisiana

You neglected to mention that the Irish in the occupied north of Ireland don't have problems with garden pests. They have no gardens because their land was taken from them by force 813 years ago.

The English pests that the Irish nationalists have to deal with are slightly more sinister than aphids and slugs: six-ton tanks, British paratroopers, RUC sadists, Orange paramilitary hoodlums, and institutionalized terrorism.

If those English suburbanites in your report cared as much about human life as they do about their tomatoes, they would insist that their government withdraw from Irish soil and at long last allow liberty and justice to take root.

Gail L. Redmond
North Arlington, New Jersey

The Police Station in Morley Safer's segment was "marigolded" not "geraniumed." Tsk! Tsk!

Ruth Castle
Oak Park, Illinois

You slipped—no self-respecting gardener refers to soil or earth as *dirt!*

Anna M. Boulanger
Lee, Massachusetts

There is not nor has there ever been a "carp" in *Plumbago capensis* nor for that matter has there ever been a "drum" at the end of *Rhododendron.*

To garden well is a mark of civilization; to check one's facts before airing them is a mark of professionalism.

Barry R. Yinger
York Haven, Pennsylvania

The plant you refer to as *Plumbago carpensis* has been referred to as *P. capensis,* named after its place of origin, the Cape of Good Hope. That however is no longer correct, since the name has been changed to *P. auriculata.*

David W. Hannings
Associate Professor
Ornamental Horticulture Department
California Polytechnic State Univ.
San Luis Obispo, California

Who cares about British flowers or weeds in the garden. I watch 60 minutes because of the funny man who always complains about things. I would like to see him every Sunday at the end of the show.

Patricia Bond
Brooklyn, New York

CHAPTER 4

People

When "60 Minutes" profiles people, celebrities or not, the mail that comes in is generally not the typical gushing fan letter. Even a beloved singer can create heated controversy. "We *all* walk the line," wrote one non-fan in disgust after "60 Minutes" rebroadcast a segment on the popular country singer Johnny Cash.

There are exceptions to this rule. Sometimes the broadcast brings to national attention somebody who is not well-known but whose achievements merit a closer look. One of these, a 76-year-old woman who is the oldest officer in the U.S. Armed Services, drew hundreds of effusive letters that ranged from proposals of marriage to suggestions she run for president.

More controversial was a two-part exclusive interview with mobster Joe Bonanno which brought in quite a different kind of mail.

And thanks to segments "60 Minutes" did on the two men regarded as the best tenors singing today, Luciano Pavarotti and Placido Domingo, we have here preserved what is possibly one of the most passionate letterwriting campaigns ever conducted.

"The Captain Is a Lady"
(Captain Grace Murray Hopper, USN)

"With [Admiral Hyman] Rickover out," Morley Safer asked the audience in March 1983, "who is the oldest officer now in active service? Well, it's not a man. The captain is a lady. She is Grace Murray Hopper, Captain, U.S. Navy, age 76, just as feisty as Rickover but without the bitter aftertaste."

Grace Hopper was also one of the pioneers in the computer industry; she helped program the very first computer, a "vacuum-tube monster" called Mark I, at Harvard. In the broadcast she was shown teaching a class how she herself learned to comprehend what a nanosecond (a billionth of a second) is. "Finally one morning in total desperation," she recalled, "I called over to the engineering building and said, 'Please cut off a nanosecond and send it over to me,' " With that, she triumphantly held up a piece of wire just under a foot long—"the maximum limiting distance that electricity can travel in a billionth of a second." Then, for comparison, she held up a "microsecond" (a millionth of a second)—a huge loop of wire, all 984 feet of it, to represent how much faster computers have become over the years.

Safer talked to the captain about her attitude toward women in the military. Saying, "She is for women, but not necessarily a feminist," Safer asked Hopper, "Do you think that women should be allowed to go in combat duty?"

"I don't see why they should," said Hopper. "If I were a man in the army or the marines, for instance, and I knew that I had a couple of women either side of me, I sure wouldn't trust either of them. . . . I'm not sure that they could stand up to it."

Hopper had lots to say about the failure of leadership in America, which she attributed largely to the influence of business schools. "It spread through the colleges . . . We had courses in business management, and it didn't emphasize leadership. . . . Back in the old days when I went into the navy one of the first things we learned was . . . leadership . . . a two-way street. It was loyalty up and loyalty down, respect for your superior and take care of your crew. And that's what we lost somewhere. I think a marine would say, 'When the going gets rough, you cannot manage a man into combat, you must lead him.' "

The segment on Grace Hopper ignited a small brushfire in living rooms across the country.

Capt. Grace Hopper should be bottled and sent to every manager, supervisor, boss in every organization in the United States. They all need her, for management, not leadership, is the name of the game these days.

Loyalty up, loyalty down, she says, is the way to get the best out of your troops. This concept is foreign to American business. The guys in the wingtips are worried about saving their own jobs and justifying their fat salaries more than anything else. They don't feel much need to concern themselves with leadership roles until there's trouble in the ranks. Their MBA's didn't teach them how to deal with the human beings who work for them.

Grace is a national treasure who could probably walk into any boardroom and have them all sitting at attention in about 5 minutes. Let's hear it for excellence.

Ms. Edie McManmon
Boston, Massachusetts

I thought you might be interested in my letter to Cpt. Hopper:
Dear Cpt. Hopper:

Your appearance on "60 Minutes" on Sunday, March 6, 1983, was very inspiring. Although you have more experience than our first woman who will go into space later this year (Sally Ride), you are obviously from the same school. I hope I have your permission to quote you occasionally, especially your one statement that "A ship in port is safe, but that is not what ships are made for." If we don't learn to take risks, we never succeed.

Ann P. Bradley
Deputy Associate Administrator for Management
National Aeronautics and Space Administration
Washington, D.C.

cc: Dr. Sally Ride

In the summer of 1981 we found a small item in the *International Herald Tribune,* which explained that a real bug had been involved in the origin of the use of the word "bug" to mean difficulty with a computer program. During her work at Harvard in 1945 Captain Hopper and her associates found that the source of a malfunction they were experiencing was a moth in the works. She said that the remains of the moth are still taped in her log book, stored in a Navy archive.

Being computer hobbyists, my husband and I clipped the story, and it's still pinned to the bulletin board next to our home computer (can't xerox it for you because it's faded and crumpled).

Are you aware, by the way, that the world's first computer programmer was a woman? She was the mathematician Ada Lovelace (daughter of the poet Byron), who worked in association with Charles Babbage.

Phyllis Cohen
New York, New York

In the past, computers have caused fear and trembling in many an outsider, now everybody wants to get into the act. Remember the kids in high school and college who stood off by themselves, the ones nobody talked with because they were different? Most of those people are now presidents of computer or electronics firms. They are the ones leading this revolution. It's called "Revenge of the Nerds."

David M. McWalters, President
Datarep, Inc.
San Jose, California

I was a long-time hold-out but I finally gave up and accepted a television set from my grand-children when I was seventy years old. So I've been watching "60 Minutes" for a decade.

Last night was the first time that I have been moved to write and thank you for a truly enjoyable 20 minutes. Here we met a well-born, well-educated woman of superior intelligence who is proud of her country, proud of her work in the armed forces, and proud of filling a definite woman's role in life. More power to her!

Lucy Starbuck
Portland, Maine

Perhaps hers is the life I had always dreamed of—but I followed the "expected and accepted" rule of marrying and having those 3 children.

I would love to kiss her cheek and say, "Congratulations sailor, a job well done." That is perhaps not possible, so I shall just say, "Much love and smooth sailing."

Mary J. Lewy
Issaquah, Washington

Thank you very much for your story on Captain Hopper. She is very refreshing and a brilliant lady. I've decided to get a degree in math so I can be an officer in the navy and serve my country.

Susan T.
Long Beach, California

I only wish that God had given me such intelligence and wit. I never before heard of a "nano-second," but I sure would like one so that I could make my grandchildren say: "Nanny isn't so out of it—she knows computer terms."

Pearl Bernard
Bristol, Connecticut

We were so impressed with last week's segment on the Naval Computer Wiz, that we think you ought to change the name of your show to 3,600,000,000,000 (3 trillion, 600 billion) nanoseconds.

Carol and Arthur Hagler
Brooklyn, New York

If we could diffuse throughout our population Captain Hopper's qualities of common sense, duty to country, and leadership by example, our need of weapons would be greatly diminished. Our national character would be our invincible strength.

Col. Waite W. Worden
U.S. Marine Corps, Retired
East Burke, Vermont

Morley Safer's interview with Grace Hopper was representative of the conversation (and arguments) we've all had with Grace over the years—a woman of her convictions, to the point, and utterly sure of herself. Those of us who know and love her wish her many many more nanoseconds of leadership.

Vico E. Henriques, President
Computer and Business Equipment Manufacturers Assoc.
Washington, D.C.

Here is a woman who made it and without ERA!
It proves those who can, do! Those who can't, cry "Sexist!"
The Bimbos of NOW should shut up and listen.

Charles E. Blythe
Dallas, Texas

Captain Hopper is one fantastic woman, but Morley Safer should look up FEMINISM in Webster's New Collegiate Dictionary '81. FEMINISM is the philosophy of political, economic, and social equality for women, and Captain Hopper lives it. Female describes a condition of birth. Feminine describes certain traits. But FEMINISM is an act of faith in humanity.

To be a feminist is to be in very good company. Some of the early feminists were Susan B. Anthony, Sojourner Truth, Molly Pitcher, and Anna Ella Carroll and Abigail Adams and Elizabeth Blackwell and Alice Paul and Anonymous.

Cindy Judd Hill
Pittsburgh, Pennsylvania

I agree with the sagacity of Captain Hopper 100%. From what I saw in landings at Guadalcanal, Tarawa, Saipan, Tinian, and later in Korea and Vietnam, I too would prefer to have women much further to the rear than on either side of me.

Col. Paul DuPre (USMC Ret.)
Port Hueneme, California

Given men's preoccupation with sexual satisfaction I wouldn't want to be one of two women in a trench with a man, either. Give me three women any time. Try a battalion of women and discover if the female of the species is indeed more deadly than the male. You men might be surprised—as well as saved from a fate worse than death!

You actually opened a very tender wound. Hundreds of WAVES were discharged during WW II for marrying, hundreds more for getting pregnant, and most of us were unable to maintain reserve status through no fault of our own. It's tough to see those men getting their pensions and benefits now. Further I assume there are thousands like myself whose husbands were happy to use (up) their wives' VA home loan benefits but wouldn't pay their NSLI (in those days wives didn't have personal incomes). And then divorced them. Did Capt. Hopper ever marry?

Carol T. Oudegeest
Sunnyvale, California

I can name you six lieutenant colonels, five colonels, and a flock of general officers that I would trade for just one year's service under the command of this leader who just happened to be a woman.

SFC George W. Gatliff
Malvern, Arkansas

The question of women fighting in war remains a feminist issue to some, but the real issue as I see it is that no young person, male or female, should be faced with war again.

Catherine Andrews
Toledo, Ohio

The lady is a captain—and just think, she made full captain *without* the ERA!

Margot Wittenberg
Carefree, Arizona

Except for her anti-woman statements, we fell in love with the charming and erudite Capt. Grace Hopper. But we have one question—Why the hell isn't she an admiral?

Sandy and Richard Pirwitz
Toledo, Ohio

Captain Hopper's loyalty to the navy is commendable, but does she realize that a *man* of her genius and eloquence would have been made an admiral years ago?

Donna and Ric Alpert
Matthews, North Carolina

It was my husband who pointed out that if Capt. Hopper were a man she would be an admiral—or, more likely, president!

Vicki Vance-Harrell
Mountain View, California

"Man of Honor"
(Joseph Bonanno)

In the spring of 1983, Mike Wallace took the opportunity to do the first full-length TV interview of Joe Bonanno, for three decades one of the leading Mafia figures in the United States and considered by many to have been the model for the title character in *The Godfather.*

Bonanno, then 78, had written a book and agreed to the interview because, said Wallace, "he wants the American people to know he was not the crime figure the law-enforcement people and the press made him out to be, but that he was, instead, a man of honor."

Bonanno was soon to go to federal prison for conspiracy to obstruct justice, but he was still at home in his lavishly decorated Tucson house when Wallace sat him down along with family members to talk. Bonanno looked handsome, silver-haired, rosy-cheeked—the picture of expensive grooming.

Though Wallace pressed tough questions on him, Bonanno turned on the charm and managed to avoid serious self-incrimination. At one point in the program, he accused Joe Kennedy, father of JFK, of being a bootlegger and a partner of crime boss Frank Costello—an allegation in Bonanno's book that had been denied by a spokesman for the Kennedys.

Bonanno's son Bill—according to law enforcement officials the *consigliere,* third in charge of the crime family—was more forthcoming and articulate. He himself was facing state and federal charges in California for conspiracy to commit mail fraud and interstate transport of stolen property.

"Are you telling me," said Wallace to Bill Bonanno, "that there is no such thing as organized crime? There is no Mafia?"

Bonanno answered, "Certainly not. Organized criminal activity is as American as apple pie. . . . The turn-of-the-century Jewish gangs, the Irish gangs, the Italian element . . . now we have the blacks, the Puerto Ricans, the Cubans, the Colombians. It's an endless stream of activity."

One of the most revealing moments in this two-part interview with the Bonanno family came when Wallace confronted Bill, the only one of Bonanno's sons who had been permitted to get involved in the family's criminal activities, about his decision to do so: "Why didn't you turn your back on it?" asked Wallace. "[You're] a man of intelligence yourself, attractive, gift of command, articulate, educated—and a hoodlum. Why?"

Bill Bonanno seemed to turn thoughtful, less defensive. "Do you know how many times I've asked myself that question? And it all comes down to one thing—love of heritage and love of another person [my father]."

A year after the interview aired, Joe Bonanno, from his prison cell, sued the publisher of his autobiography for $18 million because the cover on the paperback version had an illustration of a man in "the typical dress of a cheap gangster–professional killer."

Where would Joe Bonanno be if all the so-called legitimate people refrained from taking a drink or placing a bet in the days when such happenings were forbidden?

Our judicial system wants to label him a criminal. He did not cause harm to anyone outside his own world. Can we make the same claim about one of our presidents?

Alphonse J. Cavale
Chesapeake, Virginia

Well, Mike, after the Bonanno Family story it seems to me you have an even more interesting story about the Kennedy Family. Why don't you find out if Joe Kennedy did in fact make his fortune on the misfortunes of others?

Think of the ratings!

Daniel J. Sherwin
Olean, New York

There is nothing you can tell us about the Mafia that we don't already know except that this hood wants to sell his book on prime-time TV.

Tony Teju
Los Angeles, California

The close relationships that exist in the Italo-American family stand as a model that all other American groups might well emulate. I'm a Professor of Engineering at Penn State U., not Italian, & hope you *haven't hurt* the Italian's self image. For 30 years I happened to live in Italian neighborhoods in the Bronx, NY, & I know our country would be a lot better off if more families would follow their good example.

Quite on the other hand, an Italian family which is also connected to the Mafia is thoroughly atypical.

George Grenier
Middletown, Pennsylvania

Your program has contributed much in the past toward uncovering schemes, scams, frauds, and other unsavory & questionable practices in this country. Also you have produced other memorable and enlightening things of interest to the public.

What in hell I can't understand is why you have to publicize, glorify, give credence to, and otherwise portray Joe Bonanno as an important and worthy figure in our society. Because whether you realize it or not, this is one of the things you accomplished.

Probably Joe Bonanno never felt so good in his life as when he was being interviewed for your story. As you know, or maybe you don't know, men who lead a life of crime usually in later years like to be respectable in society. And thanks to you, he has achieved some measure of this. Now he can boast, "I've been on television all over the country." Keep up the good work: I can hear the criminals in the back rooms saying, "Let's get Mike Wallace over here and we can get on TV."

Ray Reiner
Palo Alto, California

I was surprised that Mr. Wallace, who displayed a great knowledge of the history and dealings of these powerful Italian families, did not discuss the original purposes for the father being vested with such authority.

This *Patria Potesta* harks back to Roman days and can be literally translated "Power of the Father." Many times, when state and local governments had little or no power to cope with the people's problems, the central father-figure became the tribal law.

Furthermore, in your desperate attempt to expose Mr. Bonanno and the questionable methods he used in his rise to power, you completely forgot to ask about one obvious benefit which the Bonanno family enjoys: How many of them are ever mugged, beaten, robbed, raped, or assaulted in any way?

Joseph U. Siriano
Ashville, Ohio

It seems the only Italian word that non-Italians can pronounce is pizza.

I got pasta-tively tired of hearing Mike Wallace pronounce Joe Bonanno's surname almost like banana—as in: "Yes, we have no *Musa paradisiaca.*"

It is pronounced bawn-AHN-naw, and if his ethnic background had been Anglo-Saxon, then his name would be Joe Goodyear—for that is precisely what *bon anno* means in dialect . . . and, by extension, it also means "Happy New Year" in some circles.

If he doesn't go to jail—and decides to buy a zeppelin—don't you think The Bonanno Blimp would be catchier than The Goodyear Blimp?

Italically yours,

Vivian Fiore Christy
Inwood, New York

Great Italians and their contributions are frequently overshadowed by a few people of the same ancestry. I'm sure I speak for a great many of my Italian brothers and sisters in that we prefer to focus upon the contributions made to mankind by such Italians as: Leonardo Da Vinci, Galileo Galilei, Michelangelo, Giuseppe Garibaldi, Enrico Fermi, etc., rather than the "contributions" made by Joseph Bonanno, Charles Luciano, Al Capone, Carlo Gambino, etc.

We have the right and privilege to be proud of our brothers and sisters who have so richly contributed to Italian heritage and culture. We have virtually no recourse but to feel sorry for these unfortunate and lonely men of organized crime, death, and destruction.

Rick P. Bonanno
Sheboygan, Wisconsin

I felt it was very inconsistent to say that mafioso families (Bonanno's specifically) "protect their women." It is well known that many of these families make their money through prostitution.

This Italian family was such an overblown stereotype; I felt I was watching *The Godfather Part III.*

Lisa Canadeo
Milwaukee, Wisconsin

Mike, I have seen you tougher interviewing clerks in a civil service office! I have seen you tougher on senators, but then they don't play by the same rules as Joe Bonanno. Right, Mike? No one's ever gonna call Mike Wallace stupid!

But your credibility, Mike. It's gone. You're gonna live to a ripe old age (maybe even 78), but you will never have any peace after that interview.

You blew it, Mike. You had the opportunity of a lifetime, but Joe Bonanno turned you to "mush."

Richard J. Moran
Pomona, California

Surely your network's programmers have beaten me to the punch, but here is my concept for a new TV series. It is called "Bonannos"! I see "Papa Bonanno" as the patriarch of a big "Turf," with cattle, vegetables, etc. He is widowed, but has three children—"Little Joe," "Ben," and a daughter who runs the house, cooks, cleans, and is protected from all of the "Naughty Bits" that swirl around the clan every week. Each week they face everything from the heads of government to the heads of other clans and always win when "Papa" brings out his secret weapon: "Great Respect."

Mike, you know this has got to be a "Hit"! Since you are acquainted, do you think they could be signed to a "contract"?

Mr. C. L. Woodward
Tulsa, Oklahoma

Joe Bonanno's son told Mike Wallace that "crime is as American as apple pie."

He's right. The way my American mother made apple pie was a crime.

And the syndicate (or "Mafia") is always in there for the largest portion of the pie.

Al Hamburg
Champaign, Illinois

The Great Domingo-Pavarotti Debate

On July 24, 1983, "60 Minutes" reran Morley Safer's segment on Luciano Pavarotti. The next week featured a reprise of Harry Reasoner's profile of Placido Domingo, the Spanish tenor.

In the Pavarotti broadcast Safer said, "There's been for some time a war going on, a kind of voice war, the war of the tenors. Who is the greatest? In one corner is Placido Domingo; in the other, Luciano Pavarotti. We've reported on both men, both of those spectacular voices, but I think Pavarotti wins—not only because of the voice but because of the personality. Not since Enrico Caruso has an operatic tenor so enraptured an American audience the way Pavarotti has."

The next Sunday, Reasoner prefaced his segment on Domingo by saying, "The winner, if there can be said to be a winner, is in the eye and ear of the beholder. You've already beheld one. Now behold the other. And you decide."

When you behold these letters, it will be easy to decide that there is no such thing as "summer doldrums," that cliché for TV rerun time. "Doldrums" means "a becalmed state," and that's a state most opera fans apparently have never visited.

Come now, Mr. Reasoner, I know it's cold in New York, but kindly remove your ear-muffs and listen.

Pavarotti's voice is like a fine wine aged to exquisite perfection, his charm like a cozy fire warming a log-cabin as it huddles against a snowy mountain.

Domingo? Like pizza and beer. Everyone likes pizza, that's the problem.

Jutta Metten
Los Angeles, California

Test the three candidates for "Greatest Tenor of the Century" with a delightfully simple rule:

Enrico Caruso, Luciano Pavarotti, and Placido Domingo have all recorded classical Neapolitan songs ("O Sole Mio," "Torna Surriento," etc.). Make concessions for Caruso because recording techniques were primitive in his day. Make concessions for Domingo because he's a Spaniard singing Italian folk songs. That done, hear for yourself whose voice, artistry, and musicianship come through as the greatest.

Pavarotti wins—hands down!

Carl D. Soresi
Englewood, Florida

I believe that there are two kinds of singers, those who make music a vehicle for themselves and those who make themselves a vehicle for music. Pavarotti and Caruso are special members of that wonder-working musical community. That, combined with astonishing beauty of voice and remarkable generosity and warmth of personality, have made them partners in *primus*, not to be compared with each other but to be equally treasured. I barely missed Caruso in person but I am blessed by hanging in long enough for Pavarotti.

June Edson
Montpelier, Vermont

When you showed Placido Domingo performing, he was singing baritone, not tenor!!! Caruso was a baritone who was able to reach tenor high notes in his repertoire but he was not a lyric tenor, and neither is Mr. Domingo. I like them both, but Mr. Pavarotti is *la prima voce tenore nel tutte el mundo* (Mr. Pavarotti is "the first tenor voice in all the world"). Nuff said.

Henry L. Alier
Dallas, Georgia

My letter is directed to the inquiring attitude of "60 Minutes." Hail! for an open forum where all can speak their piece if they want.

Ahem! My piece! The uniqueness of Pavarotti has nothing to do with his size, his acting ability, his adoring fans, his many curtain calls, or his devotion to his family.

What we're evaluating here are voices.

We're confronted with two tenor voices; however, only one has the tenor sound; the other has a magnificent voice with not only tenor tones but unbelievable baritone sounds.

Classical music has been my life since I was 5. Frankly, I have always disliked the tenor sound. No longer, for God has created such an instrument that, now, it is a joy to watch and listen to *Bohème.*

Thanks for letting me get this off my chest.

Jean Kennedy
Richmond, Virginia

I used to wonder what the difference was between Placido Domingo & Luciano Pavarotti. I have found the answer to this most vexing question: When I listen to Domingo, I hear a very pleasing voice. When I hear Pavarotti, I get goose pimples.

Madeleine E. Ausel
Stroudsburg, Pennsylvania

Your statement that Pavarotti has won the "battle of the tenors" is an opinion you are certainly entitled to, but let's be realistic—Have you checked to see which tenor has the most bookings? Which tenor is opening the Met's celebrated 100th Anniversary season?

Domingo performs more roles than Pavarotti and he certainly is not the temperamental, fragile-voiced tenor Pavarotti is portrayed as being. Domingo sings with greater ease—at least he doesn't have to hold a handkerchief to mop the profusion of sweat that Pavarotti produces when he performs. Placido Domingo has everything an audience could want: He's handsome, has the ability as an actor which is rare on the operatic stage, and produces sounds so gloriously beautiful that they are felt to the very depth of one's soul.

How about settling this "duel of tenors" by showing a film clip splicing the tenors side by side, singing the same aria, and switch from one to the other. Gentlemen, this will prove once and for all that Placido Domingo is the greatest tenor of our decade, and for many to come.

Hazel Fox
Albuquerque, New Mexico

Have you had your hearing tested lately?
Placido Domingo is the greatest living tenor.
Helluva soccer player too.

Margaret Larner
Philadelphia, Pennsylvania

How dare you say that Luciano Pavarotti is the No. 1 tenor today? His voice may be nearly as good as Placido Domingo's but Domingo is the professional par excellence—his voice is beautiful, his technique flawless, his musicianship perfect, he is a terrific actor, has a good personality, and in addition to all that, he is sooooo handsome! Pavarotti portly? Are you kidding? Maybe Mr. Safer needs glasses. Pavarotti is a blimp!

Stella C. Mahoney
San Antonio, Texas

I am not a letter writer but I must protest your conclusion that Luciano Pavarotti wins over Placido Domingo. Domingo's personality is more appealing. His charm and humor more delightful. But on that point we can just disagree.

However the voice lends itself better to analysis. Domingo's is warmer, more expressive, more sensuous, and capable of singing a greater range of tenor roles. Pavarotti's voice has a steely quality which causes it to be less expressive, lacking the sensuality which tenors need considering their romantic roles. There is even a certain sameness in his singing.

We will not even go into appearance, which puts them poles apart.

Barbara Schadler
Upper Montclair,
New Jersey

Mr. Safer, if you don't know what you are talking about then please don't do the story. For one thing a piano does not have to be tuned everyday, as you said, unless it is a piece of garbage that has a warped pin block and loose pins, not unlike yourself.

Chubby Luciano cannot shine Placido Domingo's shoes. You said Pavarotti performs rare songs, which is a joke. Every piece I heard in your segment was a standard, typical, and common program any and every tenor would perform—"Operatic Hits," so to speak. Please, Morley, next time just say it's your *opinion* Pavarotti is the best, which of course he is not—the music he sings is the magic, not that blimp.

Robert Laine DeVito
New York, New York

I love opera!
I love Domingo.
He is the top tenor in my book.
Send Fatso back to Italy!!!

Mrs. John Karpan
Keewatin, Minnesota

HELP!

My family and I just learned that we missed seeing and hearing our absolute favorite singer, Placido Domingo, on "60 Minutes."

We consider ourselves faithful watchers of "60 Minutes" but do occasionally go out on Sunday evening. Since our local papers do not list what will be featured on the program (a pet peeve!) we sometimes find we have missed something we wanted very much to see and for which we would have changed our plans. We are appalled to think that we didn't miss Johnny Cash but did miss Placido Domingo!

PLEASE, PLEASE, repeat this episode, and, if at all possible, please return the enclosed envelope with a note telling us when we can expect to see it.

Mrs. Ernest S. Wolfe
Long Beach, California

Compared to the great Chaliapin, Placido Domingo and Luciano Pavarotti are castrati.

W. H. von Flugen
Woodbine, Iowa

I wouldn't expect this of you. How dare you ask Domingo if he cheats on his wife?!

This is the second time "60 Minutes" has let me down.

Roger L. Kantz
Denver, Colorado

It is obvious to me that Harry Reasoner is not acquainted with any Italians or people of Italian descent.

In one previous story about Naples, he referred to pasta (pahs-ta), as past-a. I couldn't believe my ears. Now, in his story on Placido Domingo, he referred to Luciano (Lu-chee-ano) Pavarotti, as Lu-si-ano Pavarotti.

Come on Harry, you must be more cosmopolitan than that!

If you need further help in Italian pronunciation, I am a great fan of Mr. Reasoner, and I offer my assistance.

Irene Trionfetti Kay
Fort Lauderdale, Florida

"Besame Mucho" the worst song in the last fifty years, indeed! You are referring to one of the most popular songs of my, and many others', long lost youth.

Mitchell F. Stanley
Youngstown, Ohio

I don't understand these opera freaks. I couldn't understand anything the man sang. Led Zeppelin would be much nicer.

Dan Cade
Terre Haute, Indiana

CHAPTER 5

Dear Mike

Even after all these years (he's been with "60 Minutes" since its first broadcast in 1968), Mike Wallace loves getting mail. He knows exactly when the guy from the mailroom is supposed to deliver (twice a day, at 10:30 A.M. and again at 2:30 P.M.), and looks forward to it. When he returns from a trip one of the first things he likes to do is see what mail has come in during his absence.

"In the early days, I used to look at *all* the '60 Minutes' mail, because I wanted to get a sense of what the audience felt," he recalls. "And the main thing I learned was that there seemed to be an extraordinary personal relationship between viewer and correspondent. We were seen as ombudsmen; people were counting on us to take up the cudgel against malfeasance, to right wrongs, to even the balance.

"There were lots of letters that said 'Right on, Mike, keep going after the bastards.' I got a sense that people felt, perhaps for the first time on TV, that somebody *was* going after the bastards, and they were delighted."

Now, of course, neither he nor anybody else on "60 Minutes" has time to read all the mail that comes in. So he sees only mail addressed to him unless somebody in the office passes an otherwise interesting letter along. His assistant opens his mail, taking care of things like requests for transcripts and autographed pictures (all the correspondents except Andy Rooney supply them when asked). "Fan letters I give to Mike," she says, "and now and then nut mail, if I think it will amuse him."

But the mail he likes best offers program suggestions. He takes the most recent delivery and spreads it out on his desk, which it nearly covers. One large stack consists of big, overstuffed manila envelopes. They are covered with postal stamps: "Certified," "First Class," "Special Delivery." On one of them is a handwritten note: "Open before next Sunday's program."

Most of this mail contains program suggestions—he can tell by its bulk and by the official stamps, which suggest how important it is to the sender that this parcel reach its destination. He's glad to see them; he knows there's always a chance that buried within all that paper is a gem of a story.

"Some of our best pieces come out of the mail," he says. "For example, the segment on kepone, the one on the grading of meat. Maybe one out of five or six pieces I do originate in the mail. Amway was another one. Our viewers work in effect as stringers; they write to tell us a story only they know about, or send a clip from a local paper."

Intrigued, now, by the pile on his desk (he doesn't normally go through unscreened mail), he starts to look through it, glancing at return addresses on envelopes. One bears the name and address of a temple. He chuckles. "Probably accusing me of being pro-Syrian, anti-Israeli," he says. He picks up another letter and opens it; it's a request that he send some personal belonging to a celebrity auction. "Into the wastebasket," he says.

He has one letter framed and displayed in his office. It is on letterhead engraved "The Vice President/Washington." Dated February 28, 1969, the letter reads, "Dear Mike: Come have lunch with me and I'll pay you. Sincerely, Ted."

He enjoys the ambiguity in that letter, the questions it naturally raises in anybody who reads it. But it is not about shady business, unless you think betting is immoral. "Spiro Agnew, you may recall, was from Baltimore," explains Wallace. "I had bet him on a New York Jets–Baltimore Colts game and I won. And he hadn't paid me."

There are other letters that he keeps, unframed in his files, because they mean a lot to him. One came from Helen Hayes after he interviewed her in December 1983. Her letter includes what must be one of the most unusual characterizations of tough Mike Wallace ever committed to print, when she compares him to TV's most renowned and soft-spoken children's-show host:

"Together we seem to have made some mighty powerful music that could waltz us both into new careers. Me, as a regular on talk shows and you, my dear, as a new Mr. Rogers, that dear, sweet, kind fellow. You did a masterful job of making me sound better than I have ever sounded before, and I thank you. Mostly, I thank you for being so gentle, so kind, so supportive."

Wallace shows this with great pride. He also recalls, and is able to locate, a letter he once received from the man who was in charge of the U.S. military effort in Vietnam. Dated March 20, 1972, it is on letterhead imprinted with the flag of the five-star general.

"Dear Mike," it begins, "I am not one that spends a lot of time looking at television, but yesterday afternoon, after taking a shower following a tennis game, I got in on part of '60 Minutes'—fortunately, that part telling the story of several army veterans who had lost limbs in Vietnam. I just want to tell you that it was a first-class piece of reporting; I have never seen better. It told a subtle story that was both touching and realistic as to the differences in the spirit of men. I am at this seating also writing to Captain Kirk, whose courage and outlook should be inspirational to all. My compliments to you." The letter is signed, in a bold scrawl, "Westy," over the words W. C. Westmoreland, General, United States Army, Chief of Staff.

Ten years later, "Westy" has brought suit against "Dear Mike" and others over a CBS documentary program on Vietnam that he did *not* like—a suit that, as of this writing, is still pending. The lawsuit seems to pain Wallace, even though he realizes such risks go with covering controversial territory. The old letter, though, like the many he gets from "ordinary" people every day of the week, gives him pleasure.

CHAPTER 6

Guns

"Warning: Dangerous to Your Health"

In October 1982, Mike Wallace reported on the efforts of Dallas lawyer Windle Turley to make manufacturers of small concealable handguns responsible for the death and injuries that result from use of those guns, frequently called "snubbys" or "Saturday night specials."

Turley has made a national reputation by winning huge settlements from the manufacturers of defective products, and he believes, as Wallace put it, that "handgun manufacturers should be just as liable to pay victims—not because handguns don't work, but because they work too well."

Wallace reported the gruesome statistics: One-half of all murders in this country are committed with such handguns—12,000 a year. They are responsible for 350,000 injuries a year. More Americans are killed annually in the U.S. with handguns than were killed in a year in Vietnam. And all this means over a billion dollars annually to pay hospital insurance, medical welfare, and court costs.

On screen, we rewitnessed the shooting of George Wallace and Robert F. Kennedy, and heard the screams of bystanders. We were reminded that John Hinckley used a snubby to shoot Ronald Reagan. And we saw the flickering black-and-white film of a store security camera as it recorded the death of an unnamed store owner, shot by a snubby as he stood by his cash register.

Wallace interviewed the sheriff of Florida's Broward County, who said, "I personally believe that the handgun has no place in self-defense." As he noted how often handguns are used not by strangers or criminals but within law-abiding families, newspaper headlines were flashed on the screen: "Man Accidentally Kills Friend," "Family Feud Ends in Death," "Boy Kills Self with 'Unloaded' Gun."

The manufacturers of four of the handguns most often used in crime in the United States declined to talk with "60 Minutes," as did the National Rifle Association. They suggested Wallace talk to Bill Ruger, a top gun manufacturer who, however, does not make snubbys. Ruger, sitting in his office in front of an old-fashioned rolltop desk near a display cabinet filled with rifles, said, "This Turley business, they're not the least bit interested in drawing a line anywhere. They'll escalate that thing right down to the point where every shotgun in America has become the subject of a lawsuit."

As always when the subject of gun control is raised, this segment drew considerable mail from NRA members. Much of it was peppered with the bumper-sticker slogans popular with some gun owners: "Guns don't kill people, people kill people." "When guns are outlawed, only outlaws will have guns." But there were also many letters from people outraged by the easy availability of guns in America.

Wallace reported at the end of the segment, "Windle Turley has apparently triggered a spate of similar lawsuits, among them that of former presidential press secretary James Brady, whose attorneys have now filed a multimillion-dollar suit against the RG Gun Corporation that manufacture and assemble snubbys like the one John Hinckley used."

Your airing of the opinion that anyone wishing to legally carry a concealable handgun does so with the desire to commit criminal acts, makes a white handgun owner like myself aware of what the word "Nigger" must mean to a black person.

I'm not a member of the NRA now, but will be by the time you get this.

Richard F. Way
Reading, Pennsylvania

As your segment, "Warning: Dangerous to Your Health" (October 24) proved beyond a shadow of a doubt, small concealable handguns are a criminal's best friend and a threat to the very fabric of our society. It's no wonder that those who manufacture these instruments of death and their pimps, the NRA, refused to speak to you. By using legally fallacious interpretations of the Second Amendment and scare tactics, they have contributed to the bloodshed that envelops us.

Dan O'Neill
Los Angeles, California

The notion "Guns don't kill—people do" is absurd, along with the idea that violent people would simply reach for other weapons if handguns were not available. Imagine for a moment John Hinckley, Jr. springing from the crowd wielding a baseball bat, or some punk trying to hold up heavyweight champ Larry Holmes with a knife. Or Sirhan Sirhan using a blackjack to deprive us all of Bobby Kennedy before being subdued. Nonsense!

The fact is, a handgun gives every mental defective with 50 bucks (about what Hinckley paid for his gun) the opportunity not only to change your life and mine drastically, but to negate the votes of millions of Americans and change history. Let's make carrying such weapons outside the home a crime!

William R. Stermer
Oxnard, California

Last spring my happily married sister began to behave unusually. At first, she would accuse my brother-in-law of different things that upset her (which had never happened). They went to a doctor and medication was prescribed. Unknown to her family, she stopped taking it. Within weeks she would become violently angry with very little provocation, and once was disarmed of a kitchen knife when she attempted to attack my brother-in-law.

Her one obsession was to get a gun. They turned her down. Thank God!! Before this "illness" my sister was a calm, happy, loving mother. The furthest thing from her mind would have been harming any living thing. While in this violent frame of mind, her one obsession was to kill her husband. Eventually she became a patient in a psychiatric hospital and received help. Today she is back with her family, and to all appearances the same person we all knew and loved before this bizarre incident.

I thank God every day that the police in Canada have controls on who is able to qualify for a handgun. The doctor suggested the diagnosis for my sister's illness was schizophrenia—the same diagnosis suggested for John Hinckley. How fast would the laws in the United States change if someone in a position of authority had such a close relative, especially if their own life was endangered?

Name withheld
Hamilton, Ontario

Canada, with gun control laws, had only about 52 murders last year involving firearms; Britain had only about 74. Your country had 12,000. When the myopic NRA claims that "handguns don't kill people, people do"—I counter by saying, "People build houses, not hammers. But without hammers, you can't build a house."

Douglas Baker
Victoria, B.C., Canada

Your segment on handguns has caused at least one man to give up his gun forever, preventing him from accidentally or willingly doing harm to himself or someone else in the future. Keep up the good work.

[The writer enclosed the following item from the "Police Reports" column of her local newspaper, *The Olympian*, Oct. 25, 1982.]

A 24-year-old West Bay Drive resident walked into the Olympia Police Department at 9:05 P.M. Sunday and gave jailer Ed Mehan a .25-caliber pistol. The man said he had just watched a segment about handguns on the CBS television show "60 Minutes" and wanted police to dispose of his gun. The man, who still has a concealed weapon permit, said he never wanted to see the gun again.

Darlene Homann
Lacey, Washington

I nearly swallowed my cigar as I heard that red-headed lawyer from Texas say gun companies are only out for the money.

Aren't lawyers like "Mr. Esquire" from Texas also out for the money? Our country will become really civilized when we learn to do without both guns and lawyers.

Cyril E. Sagan
Volant, Pennsylvania

Statistics that purport to demonstrate that guns are more likely to be used criminally than for a legitimate self-defense are seriously flawed because of a skewed data base. Criminal offenses with a gun are much more likely to be reported to the authorities than self-defense incidents. Most people, particularly those in high crime rate urban areas, shy way from the police, and are entirely content to not report the use of a weapon in self-defense. They don't want to become the subject of a police report or possibly be found in technical violation of the law for some unwitting transgression. Self-defense incidents, with or without a weapon, are generally reported to the authorities only when an injury or body requires an explanation. Most self-defense usage of handguns does not involve shooting; just demonstrating that a handgun is present and could be used is sufficient to deter most criminals.

Handguns represent the only real defense many in our society have. The poor and unfortunate are concentrated in decaying urban centers where "police protection" is a myth. Denying the poor the only effective form of personal protection available to them is elitist. I was not surprised that the lawyer advocating gun restriction on your program drove a Ferrari; he can afford armed guards to assure his personal safety. The poor don't share his enviable resources.

John D. Dingell III
Trenton, Michigan

As much sympathy as I have for James Brady, I find it ironic that John Hinckley, Jr. should be found not guilty by reason of insanity, and yet the manufacturer of the pistol he used in the assassination attempt is facing a lawsuit for producing the weapon. Using that logic, it seems to me that Hinckley's parents should be held liable for producing him!

Darcey Snow
Ontario, Oregon

What this country needs less of is greedy lawyers who just want money, and we need tougher laws to stop criminals, not handguns!

As an American gun owner of 14 years of age, I will give up my gun when they pry my cold, dead fingers from it!

Jeff S. Brantley
Yorktown, Indiana

I am 13 years old and would like to comment on the part about the guns. To me all guns are disgusting, but the snubby handgun is one which I would like to speak out against. I've heard too many times of someone getting shot by a concealable handgun.

I am glad that there are people like that attorney who try to rid us of these terrible weapons.

Britta Cornils
Russellville, Tennessee

My brother is seven years old and I have three other brothers younger than that. My seven-year-old brother receives *Boys Life* magazine and its scouting section advertises guns and rifles; it also states places to buy guns.

I found out the ads for guns were intended for Boy Scouts of higher rank. But what about my brothers? If my five-year-old brother wanted to he could get a booklet and order form about guns. Furthermore *Boys Life* was sold in school for fund raising. I have no idea how many other boys subscribed to *Boys Life.* What is your opinion?

Katie Powers
Oak Park, Illinois

In regard to your story on Handguns. If I want to write I buy a pen. If I want to eat I buy a fork. If I want to drive I buy a car. If I want to keep food cold I buy a fridge.

Why would I want to buy a snub-nosed handgun? To shoot at Tibetan Yaks? Surely not. There is only one reason I buy a snub-nosed handgun. Can you guess?

Take away my purchase and you take away my reason.

R. J. Antosko
Beaconsfield, Quebec, Canada

I own a .357 magnum and at a cost of over $500.00 it is one of my most treasured possessions. With its easy accessibility, accuracy, and potential for being lethal, I believe it would be very effective for protection. Outlawing handguns will only cause more lawbreakers (myself included) and a black market for firearms. There are many laws restricting narcotics, but look at the growing number of drug abusers. Do the laws work?

I do, however, agree with one point you attempted to make. The handgun manufacturers should be socially responsible. But instead of doling out millions in lawsuits, they should contribute funds to support handgun safety programs.

Kevin W. Sutton
Cape Girardeau, Missouri

When you read in the paper about the break-in of a car you will often find that a gun which the good citizen had in his glove compartment has been stolen by the not-so-good citizen. The same goes for break-ins of unoccupied homes.

Enough is enough.

Angelika Ertmaier
Miami, Florida

Let's ban handguns to eliminate murders, then seize all automobiles to curb highway deaths, confiscate all cameras and printing presses to eradicate pornography—and just wait until you hear my answer to the rape problem.

Hank Babbitt
Sault Ste. Marie, Michigan

If the gun manufacturer is responsible for shootings, then Dan Rather should be jailed for the copycat poisonings following the Tylenol incident—having put the idea into sick-heads across the country with irresponsible night-after-night repeats of what obviously was a local incident.

Shim Zangara
New Albany, Indiana

I believe it's important to point out that this type of litigation would not be limited to just the small handgun. After all, the difference between a short-barrel handgun and one with a longer barrel is a 25-cent hacksaw.

If successful, these lawsuits would establish legal precedence and could be used by any other attorney to also ban the manufacture of long guns under the legal doctrine of *stare decisis.*

The goal here is obvious. Silence the majority of Americans who are opposed to the banning of firearms by going through the courts.

Michael J. Slavonic, Jr.
Legal Affairs Director
Pennsylvania Rifle and Pistol Assn.
Pittsburgh, Pennsylvania

The money raised by the anti-gun extremists, retainers for their legal counsel, criminal fines, money for anti-gun propaganda, and the cost of "60 Minutes" air time could be better contributed to enlarging existing prison facilities. Then we competent, law-abiding gun owners and manufacturers could be left to the peaceful pursuit of our various sports. This is in keeping with our Constitutional Guarantees and the wishes of our founding fathers.

Stephen R. Burgess
Reno, Nevada

The legal attack on gun manufacturers you described is the most exciting potential antidote I've heard about to stop the contagion of killing by concealable handguns. I'm a physician who helps rehabilitate gunshot victims, but I'd rather let lawyers handle prophylaxis and get me out of this particularly tragic enterprise.

Bruce H. Dobkin, M.D.
Inglewood, California

The gunshop owner: "It's not me who is responsible for a murder with a handgun; it's the owner of the gun."

The NRA adherent: "It's not guns that kill, it's people that kill."

The gun production owner: "We make guns, we don't commit murders with the products."

Please allow for a short analogy: An entire bureaucracy of thousands rejected responsibility for the death of millions in Nazi concentration camps. It wasn't the Nazi elite, or the camp commandant, or the SS officer, or even the Kapo who were responsible for the tragedies of the holocaust! In trial after trial all disclaimed responsibility.

May I suggest that the true test of a society is what future generations will make of that society. America may turn out to be the greatest society which lasted the shortest amount of time, if it doesn't take responsibility for the insanity of its ruthless deployment of all firearms.

Norman Barrie
Merrick, New York

CHAPTER 7

Pollution

Contemporary concern over environmental issues is at least as old as Rachel Carson's 1962 bestseller, *Silent Spring.* But the issues remain complicated and controversial. At a 1983 press conference President Reagan accused organized environmentalists of occasional "extremism" and joked, "I don't think they'll be happy until the White House looks like a bird's nest."

Among the "60 Minutes" audience, there are those who agree with that view. But the mail reveals many others, not lobbyists, who have become upset in the extreme when their own water supply has been threatened, or the air they breathe has begun to make them ill.

"The Spraying of Moundville"

Hundreds of people wrote in after "The Spraying of Moundville" aired on March 13, 1983. Ed Bradley had gone to Moundville, Alabama to tell the story of what happened after an increasingly common event: the spraying of utility lines with herbicides to kill excess foliage. This goes on all over the country, not only on power lines but on railroad right-of-ways and other places where leaves and vines and branches encroach on the equipment that brings us such necessities as electricity, telephones, and cable TV.

Usually, the spraying comes and goes—stinking, maybe,

and annoying a few people, but soon forgotten. But in Moundville, when the Alabama Power Company sprayed Tordon 101, an herbicide produced by Dow Chemical Company, some people got sick. And a seven-year-old boy named Randall Cephus died less than three weeks after the spraying. He had eaten an apple from a tree in the backyard of his grandparents' house. His parents say that tree had been sprayed with Tordon 101.

Said Bradley, "An autopsy report put the cause of Randall's death as consistent with epilepsy, but his parents say Randall never had epilepsy." A doctor from Alabama's Poison Control Center said on camera that after postmortem analyses of the child's body as well as apples from the tree, the center officially concluded that Randy did not die from Tordon.

BRADLEY: What did he die from?

DR. BRONSTEIN: I don't think we know what Randy died from. I don't think we'll ever know what Randy died from.

BRADLEY: Why not?

DR. BRONSTEIN: It happens every day.

BRADLEY: Really? Someone dies and you don't know why they die?

DR. BRONSTEIN: I work many hours in the emergency department, and people will come in, and they will just die, and I know this is 1983 and we've got the CT Scanner and the new NMR scanner and all this sophisticated equipment, but I can tell you that people do just up and expire."

At the conclusion of the segment, Bradley summed it up: "So that's what happened in Moundville in the summer of '82. It started with the power company attempting to clear its lines by spraying them with a chemical. It ended with a lot of people getting sick and with the death of a child. Nobody has scientific proof of any connection. And the power company says it will spray again when necessary."

Hundreds of letters came in from all over the country,

from experts amateur and professional, from people whose own lives had been touched by similar experiences—giving a fascinating glimpse of the spectrum of rage, confusion, and uncertainty we feel when confronted with the miracles and attendant mysteries that science has brought us.

What is going on here? I mean we've had enough already. First all the EPA problems and your continuing sagas of toxic waste sites, but now a modern Agent Orange (a previous story)!! This little town has a few people get sick and a little boy dies at the same time and you make a federal case of it? Are you running out of ideas? C'mon, leave the environment alone and give us some *real* issues.

Jeff Greenbaum
New Canaan, Connecticut

I hold a doctorate degree in toxicology, I am a Board Certified Toxicologist and I have about 10 years of postdoctorate experience as an expert in the toxicology of pesticides.

The factual evidence you presented clearly support the conclusion that a young boy very regrettably died of unknown causes. There was no evidence to show that Tordon was implicated in the death of the child. CBS has created controversy concerning the safety of a pesticide that has no basis in fact.

I recognize, Mr. Bradley, that you cannot be expert in all the stories that you report. Therefore, you must rely on experts. Your technical support in preparing this piece was either misguided or stupid. Furthermore, you wasted valuable time that could have been directed to addressing some of the pressing environmental problems.

M.W. Sauerhoff, PhD, DABT
Stauffer Chemical Co.
Farmington, Connecticut

Who is Alabama Power Company trying to fool? I spray Dow Chemical Tordon 101R on my hard tree stumps and in a matter of time they turn to pulp. It says on the container, "Do not store near Food or Feed." I'll spray an apple, then I'll dare anyone from Alabama Power & Electric Company to eat it!

Raymond C. Bottles
Grafton, Ohio

I've been a farmer for several years. It seems all the chemical companies do is recommend more herbicides, or stronger ones. When it comes to the chemical companies, you can get about anybody to say something in their favor. It's about time we put a stop to all of this!

Darrell Felton
Liberty, Indiana

Due to interest generated by your story on Tordon 101 and our own curiosity, we researched the Dow product line of Tordon products listed in the 1982 agricultural chemical label book, 6th edition. Of the products listed in the text under the name Tordon 101 (Forestry Herbicides) not one of them advised application by spray. In fact, the use precautions of the product in question specifically state that the applicant should "not allow careless application of spray drift," as would be the case in aerial application. Taking this into consideration, along with the fact that Tordon 101 is "harmful or fatal if swallowed and causes eye and skin irritation," it is our feeling that the fault lies with the applicant. Tordon 101 is to be applied by cut surface treatment, meaning the herbicide is to be injected or applied directly to the cut tree stump. We suggest that Alabama Power Company applicator's license be reviewed along with their method of application in this case.

Michael Kerlin and
Timothy Martin
Princeton, New Jersey

Hardly a day goes by without a story in the news about another chemical poisoning . . . yet the medical establishment is still refusing to recognize it as a health problem. Amazing!

Did these doctors get their education at Dow Chemical Company?

June Murphy
Elgin, Arizona

At one time it was customary for a civil engineer to demonstrate his faith in his ability to design a sewage treatment plant by drinking from the discharge leaving the plant and living comfortably for a reasonable time thereafter.

Those doctors and power company officials who believe that hysteria is the reason that people get sick after a spraying would convince me if they and their families would demonstrate their faith in this belief, by spending a day picnicking in a clearing during a spraying operation.

Pat J. Quinn, P. Eng
Mississauga, Ontario,
Canada

My family and I live in a rural county of Pennsylvania. Our electric company is a rural cooperative called Claverack. Each year Claverack does aerial as well as ground spraying of its power lines. In the summer of 1980, my wife was four months pregnant with our third child. It was also at this time that Claverack sprayed our power line, which comes within 100–150 feet of our house. Many summer nights with our windows open, my wife and I lay awake breathing in the fumes of the spray, which would cause our noses and throats to burn. The fumes forced us to close all windows and suffer through the heat of the house which would seem to make the fumes worse.

During that same summer, my wife would enjoy taking walks down our country road in search of wild raspberries. On this one particular day, she happened upon a bunch of raspberries that, unknown to her, were sprayed with Tordon 22. She ate a couple before she caught smell of the fumes. The next day we drove to the bushes which were now brown and withered. In the enclosed directions from a package of Tordon 22, it states that a treated area should be marked or staked off. This area was not!

On November 5, 1980, my wife gave birth to our third child, Derrick. Derrick was born with a form of spina bifida called meningomyocele. Like your townspeople and doctors in Alabama, there is no "proof" the spray's fumes or the eating of the sprayed raspberries caused this birth defect. But what are people like us to do? We are fighting "city hall." Your show makes people like us feel we have someone with "clout" on our side. Please continue to expose these terrible agencies along with their practices.

Dennis Newhard
Hallstead, Pennsylvania

In my 11 years of experience as a Registered Nurse I have known of 80+-year-old people who "up and expire" usually due to underlying cardiac or cerebrovascular disease. I have never known of a healthy 7-year-old child who "up and expired" for no apparent reason. I find the blasé attitude of the Alabama poison control center physician toward the death of this child and the morbidity of this population group intolerable.

Patty Martinell, RN, MN, CFNP
Nurse/consultant
Atlanta, Georgia

Prompted by your program on the spraying of herbicides by Alabama Power Company, I would like to point out several things that might be of interest to you.

First, there is a basic assumption here that all of the people in this town have the same type of immune system and will respond in the same way. This is not true. There is always a small group of the population who have immune systems that respond allergically to very small amounts of those things to which they had previously become allergic. This includes the herbicides, insecticides, and other chemicals as well as foods, dust, and pollens. In other words, the allergic patient requires only small amounts of chemicals to have a reaction where the nonallergic person requires large amounts—the so-called toxic dose with which toxicologists are quite familiar.

The physicians involved were looking for toxic dose reactions. Because they did not find any, they assumed that the people were suffering from "hysteria." They failed, however, to review the very adequate medical literature available that discusses problems of this kind. Dr. Theron Randolph in Chicago and Dr. William Rea in Dallas, Texas have had extensive experience with allergic responses to the herbicides, and would have much to say on the subject that would clarify the thinking involved. Dr. Randolph's book on *Chemical Susceptibility* was written in the early 1950s but apparently was not examined by the physicians. Epileptic seizures from allergies have been well detailed in a book titled, *Allergy of the Nervous System,* edited by Frederick Speer, MD, published by Charles Thomas in 1970. Even deaths have been described on an allergic basis due to the swelling of the brain itself. So certainly it is possible for a boy to have a seizure and die if he were reacting in an allergic fashion.

Herbicide examination of the apples may well have been made after rain had diluted the spray on the surfaces, but in any event the measurement of herbicides must be done very carefully, and there are only a few laboratories around the United States that do it effectively, as far as I know.

Another point that must be made concerns the family that moved out of their house. While visiting Dr. Rea this last year, I saw patients who were being treated at his facility who had similar experiences from pesticides sprayed inside their homes. Their immune systems had been overwhelmed, and after this they began developing allergies to other chemicals and even foods. This "spreading" phenomenon has been well described in the literature. As a matter of fact, Dr. Rea himself was formerly a cardiovascular surgeon, but had to give up that part of his profession in the operating room because of chemical allergies to anesthetics.

The problem of chemical allergy is increasingly common. Those who work in clinical ecology and deal with allergic patients are seeing it more frequently than in the past. Many reputable physicians who have studied and reported on this for years look upon the rapid increase in the use of chemicals as a time bomb. While most people are worrying about the dangers of nuclear power plants and nuclear waste, the more widespread hazard of chemical additions to our foods and environment continues at a rapid pace. The increasing number of reactions occurring in the minority of allergic patients is very likely a "bellweather" for the rest of the population. T. S. Eliot was probably correct that our world will end "not with a bang but a whimper."

P. John Hagan, MD
The ENT Surgical Group, PC
Wilkes-Barre, Pennsylvania

In 1978 I was to be the pilot on a United States Forest Service herbicide spray project in eastern Washington State. The frauds, hypochondriacs, ambulance chasers, and marijuana growers came forth with the usual ailments and injuries.

However, all of them experienced miraculous recoveries when informed that the project had been canceled before it started, and that the helicopter was still at base over 120 miles away.

In fairness to those of us who earn our living using pesticides, please do an investigative report on the propaganda campaign of various environmental groups opposed to herbicides, and their connection with the marijuana industry.

Steven Goodman
Auburn, Washington

If I lived in Moundville and was told that the toxic spraying would continue, my monthly checks to the power company would discontinue.

Jack Stevenson, Jr.
Winston-Salem,
North Carolina

As I watched "60 Minutes" tonight I was taken back more than thirty years. To a farm in Mississippi where I lived with my husband and children.

Our principal cash crop was cotton. The boll weevils and boll worms had to be controlled by spraying the fields of cotton with an insecticide. Not by plane but by pulling a large attachment with a big tractor.

On the morning my husband sprayed the field closest to our house I was the only one at the house. The children were working in another field farther away.

The odor of the spray was strong and it made my eyes sting and my throat burn. A few days later I began to feel very tired, no energy. I had severe headaches and became so nauseous I could not keep anything down. I was hospitalized for several days. The doctors could not find what caused my illness. I got better and came home. But only for a few days, when symptoms returned. I was again hospitalized for several days. And again, improved, returned home. We believed the spray caused my illness.

Several weeks passed before I felt well again. I remained very sensitive to spray of any kind. We began to notice the cotton plants in the field closest to the house put on misshapen bolls and the leaves grew in a very strange, deformed way. The company that made the spray was contacted. They sent a man out to inspect the cotton fields. They then inspected the drums the spray came in. It was analysed and found to be not insecticide as stated on them but weed killer.

The company paid us for what was a normal cotton yield as this spray caused a "sort of plant cancer," as the inspector said. The little cotton it made was not of good quality.

My husband died in 1976. Lou Gehrig's disease (ATS). Two years later I became ill. Lymphoma, cancer of the lymph glands. I underwent painful numerous tests. Then chemotherapy over many months.

Then I developed a skin cancer on my hand which was removed surgically. I have no symptoms of cancer now. I have periodic checks at a clinic. If I get through the rest of this year safely the lymphoma will be considered cured.

I know this has been an awfully long letter but I hope you have read it carefully.

I believe my husband's illness and death as well as my own illness was a result of inhaling that spray all those years ago.

Marguerite B. Glasgow
Ormond Beach, Florida

I have a solution for the problem at Moundville, Alabama. Instead of the power company taking even a slight chance of poisoning someone with a herbicide, why not hire people to cut down the underbrush, thus getting rid of the underbrush and helping out the unemployment situation at the same time?

I am 15 years old and it only seems like common sense to me. People in government are being paid millions of dollars to find solutions to the environmental and unemployment problems. If there are openings available, I do need to earn some extra money during my summer vacation.

Tim Brockway
Quincy, Michigan

"Chlordane"

In the chemical lexicon the word chlordane is probably familiar to many more Americans than is Tordon. Chlordane is the most commonly used antitermite treatment. For years, homeowners and homebuyers have been warned of the dangers of termite infestation; termites, it is said, once troublesome only in Southern states, are now prevalent as far north as Alaska. Few people who own homes—usually their single most valuable investment—relish the thought of seeing the piano suddenly fall through the parlor floor as a result of years of silent termite damage.

On April 10, 1983, Mike Wallace looked into the case of a Massachusetts couple who sued the manufacturer of chlordane, Velsicol Chemical Corporation, when their 11-month-old son became violently ill a few weeks after the father spread chlordane around the house to get rid of ants. A jury awarded them half a million dollars.

In 1974 the Environmental Protection Agency had banned all uses of chlordane, with one exception: treatment applied *underground* for fighting termites. However, the chemical has continued to be available at hardware stores for both amateurs and professionals. Said Wallace, "Instructions say they shouldn't, but perhaps because the label says 'spray,' they do. And, it turns out, so do some professional exterminators." He then talked to a couple whose home was sprayed with chlordane. In their case, a jury convicted the exterminating company and the judge ordered the company to train their workers better.

As Wallace pointed out, definitive evidence that chlordane is carcinogenic is lacking. He interviewed Dr. Peter Infante, a cancer specialist with the U.S. Occupational Safety and Health Administration, who believes that, based on tests on mice showing massive doses of chlordane to cause liver cancer, "we should presume that this substance may cause cancer to humans."

Wallace summed up the problem: "Few things are absolutely safe. 'Safe' really means what risk you're willing to take. The risk of not using chlordane is that many people's most precious investment, their home, may simply rot away. If chlordane is banned, homeowners, according to the National Pest Control Association, will pay much more money for much less protection, because substitutes for chlordane are about four times as expensive and give protection for only a quarter of the time that chlordane does."

Perhaps we are more frightened of insects than of vegetation (and hence more tolerant of pesticides than herbicides), for the letters that come in on "Chlordane" seemed a bit less outraged than those on "The Spraying of Moundville." Perhaps it is simply that chlordane is so much more familiar, and has been around for so long. Among the hundreds of letters, there were none describing a day the piano fell through the parlor floor. Perhaps we always overestimated the threat of termites; then again, perhaps we have successfully defeated them with pesticides over the years.

I worked for 28 years in Velsicol's Marshall plant, and have used Chlordane at full strength. I have eaten many meals around the material, also smelling the vapors off of it. My wife use to wash our floor every 2 weeks with it in water. We had 5 children (all grown & married), also 6 grandchildren, and pets over the past 30 years, and all have been exposed to Chlordane. Not the usual 2% but 100% (full strength). Velsicol may not be a "Saint," but I'd use it anytime.

Ira D. Monk
West Union, Illinois

About your commentary on chlordane, Tut! Tut! Tut! Hysteria, hysteria. . . . I have an idea. Why don't we bury all the bodies resulting from chlordane next to those resulting from the now banned DDT? By the way, where are they?

Linda J. Bohannon
Alexandria, Louisiana

Your Chlordane story would not have been needed if builders would properly construct homes. Termites can not work without moisture at a certain temperature and contact with the earth or soil. Also, anything sprayed or treated with creosote bars them forever. Another common item, burned motor oil, stops them cold. You can't eliminate termites any more than ants. They are in the earth all over, but if the construction people build properly you are perfectly safe from damage, even with the ground full of termites. I know, because I went through this because of poor construction, yet sealed them off completely.

John Kuett
Point Pleasant, New Jersey

Your story on Chlordane left out one important reason why so much of it is used. If you are selling your home and the buyer is obtaining an FHA-backed mortgage, the seller must have a certificate stating that there are no termites in it. In our area exterminators will no longer give the certification unless they can also treat it. If the average home is sold every 5 to 7 years it will be treated a great many times during its life span.

Bill Mollenhauer
Pitman, New Jersey

Regarding the application of Chlordane into the ground around the house, you failed to question one of the most likely and hazardous results. I refer to the seepage into ground water, which flows to streams, then to lakes that supply drinking water—or deeper into the ground, hence to drinking water which is pumped from wells.

Water purification systems kill bacteria; they don't remove the toxic chemicals which are dumped into our environment!

Preston W. Durkee
Livonia, New York

Being in the business of selling pesticides to the professional pest-control operators, I stand to see increased profits from the sale of Chlordane substitutes. However, as an informed consumer I would be outraged to see the product of choice, with an outstanding safety record for over 30 years of commercial and private application, become unavailable to the professional applicator.

I feel your report touched too lightly on the health statistics for the commercial applicators of Chlordane. Certainly no consumer will be exposed in a lifetime to the amount of chemical encountered daily by the service technician applying termiticides for a living. Those consumers should be made aware that studies show that applicators working with the chemical daily show no adverse health effects, and no increased incidence of cancer. The consumer should also consider that the threat of cancer as demonstrated in mice cannot be duplicated in other animals.

As consumers we already bear the financial burden of increased cost due to strict regulation and testing. I feel a deep sense of outrage at the likelihood that we will also be asked to pay for the paranoia and fear generated from sketchy and overly sensational reports which disregard the great weight of scientific and practical evidence with Chlordane; a proven, effective, and safe material.

Dennis Merrill
Los Gatos, California

So our homes are our most precious possession. Oh, really! I tend to place greater value on my personal health and the health of those I love. Do I have misplaced values? I think not!

Gerald L. Ross
Everett, Washington

I am one of those bothered by Chlordane, by formaldehyde, by monosodium glutamate, and a lot of other products which don't bother some. Exposure to formaldehyde for half an hour in a closed room brings on breathing problems that could stop my breathing if I continued to be exposed to it. Chlordane can put me in jeopardy also. MSG is less drastic in its effect, but not good for me. Some would say, I'm sure, that those of us who have trouble with these chemicals are just sickly—the exceptions. But are we the exceptions or do we have an early warning system which keeps us away from problems which will strike others later? What price will we pay for ignoring some pretty obvious effects of so-called "harmless" chemicals—and what will the manufacturers say after it is too late? Will they be ready to buy ruined homes and repair ruined lives? How hard are those who regulate these things really trying to investigate, test, and regulate?

Keep on investigating.

Patricia Neal
Huntington Beach,
California

I have a chemistry degree from the University of Florida. I have worked in pollution control with respect to water projects for the past four years.

Your point was that a consumer must balance the "known or unknown" risk of Chlordane poisoning with the risk of having his home "eaten up." This is true today. However, it is fast becoming a moot point.

Chlordane and other chemicals used to kill pests and termites have created a "super termite" just as antibiotics have created a strain of "super gonorrhea."

Scientists must alter their approach to "bug" control. They must start controlling pests by attacking the genetic code and making "bugs" incapable of reproduction. PERIOD.

The manner in which bugs may be introduced to exposure of the control agent may be based upon the "carrier" concept. Some termites, for instance, should be exposed to the pest control agent and then be placed near termite-infested areas. The termite carriers will then infect the host colony. Furthermore, other termite colonies will become infected during the time of year when termites "take to the wing."

Until this approach is applied across the board to alleviate "unwanted life form" problems, "60 Minutes" will have many choices of stories like this one on which to report.

Check out the *Miami Herald* "morgue" for documentation on the "super termite."

David Campo
Key West, Florida

There's an old burlesque routine where the comedian tries to throw something away and the thing sticks to his fingers. I was headed for the dumpster with a bottle of Chlordane when it occurred to me that I can't contaminate the garbage, the garbage men, and the landfill. Don't you think that a large number of people, upon seeing your segment, would want to throw away the stuff, and don't you think that you have some moral responsibility for anticipating that trip to the dump and for giving simple directions for the responsible disposal of Chlordane?

John F. Deethardt
Lubbock, Texas

This is a question America wants answered: why can't we develop a strain of mice that can stand up to cancer?

Jerry Rainey
Charlottesville, Virginia

Your segment on Chlordane seemed slanted. Most of the people you interviewed at their homes spoke of the *improper* use of this product. Come on, be realistic. No one would say typewriters are dangerous. If Andy Rooney were to drop a typewriter from a tall building onto a crowded sidewalk it would not prove typewriters are dangerous. It would just prove that you shouldn't trust Andy Rooney with a typewriter.

Steve Moyer
Horton, Missouri

CHAPTER 8

Doctors

Doctors and nurses have at their command equipment and techniques that would have been unthinkable twenty years ago. As a result, both the medical profession and the rest of us face moral and ethical dilemmas that once were just as unthinkable.

Scarcely a week goes by that does not bring us, through newspaper or television, yet another strange twist in this ongoing drama: Somebody wants to be taken off a respirator and allowed to die, another sues a hospital for failing to use every available method to save a loved one's life.

It seems the time has passed when we can shake our head in sorrow and say of every death, "It was God's will." With increased control over life and death comes controversy: Do you really know how you would feel if it were *your* turn to make the decision about how worthwhile it is to save a life, even your own? It is a question many viewers pondered after seeing two recent "60 Minutes" segments.

"Charged with Murder"

Clarence Herbert, a fifty-five-year-old security guard, was in a hospital recovery room following routine surgery when suddenly his heart and breathing stopped for unknown reasons. He was revived, but there had been severe damage to his brain from lack of oxygen. After watching him for three days, the doctors decided he had no chance of recovery. Before his operation, he told his wife that if anything should happen to him, he did not want to be kept alive artificially. So, at the request of his family, he was taken off the respirator. A few days later, intravenous feeding was stopped. Finally, eleven days after the operation, Clarence Herbert died.

In this segment, aired May 8, 1983, Mike Wallace reported the strange events that followed:

The supervising nurse on Mrs. Herbert's floor, who said on "60 Minutes" that she had "never seen any patient receive [such] undertreatment," reported the case to the health department, which passed the information on to the district attorney. A year and a half later, two doctors—internist Dr. Neal Barber and surgeon Dr. Robert Nejdl—were charged with premeditated murder, despite the fact that they had acted upon the wishes of Mrs. Herbert, who signed a paper requesting withdrawal of all life supports. All charges against both doctors were dropped by an appellate court.

Eventually Mrs. Herbert herself sued as well, bringing a $25 million suit against the two doctors and the hospital on, among other things, the unprecedented charge of "wrongful termination of life support."

Among the many who wrote to "60 Minutes" after this broadcast were a number of doctors and nurses, revealing their own very human concerns.

I am 71 years old and the nightmare of continued existence after brain damage is for me the one black cloud of being a senior citizen. I have a "living will" stating my wishes, but the Maryland legislature again this year turned down the Natural Death Act which would have given it legal status and protected doctors such as Dr. Barber from lawsuits.

At present our only recourse in Maryland is to avoid being

hooked up to the machines in the first place. For that reason I wear an ID tag, stating in large letters: "POSITIVELY NO RESUSCITATION, NO IV, NO INJECTIONS, NO INTUBATION," in case I am not conscious to say it for myself. We will all die, and if I can go quickly, sparing my family emotional and financial stress, and saving the medical care for younger people with far more chance of recovery, then I will count it as a final blessing in a full life.

I strongly recommend the readable small book *Let the Patient Decide* by Louis Shattuck Baer, MD (Westminster Press, Philadelphia) for anyone interested in the subject.

Dorothy Graham
Cockeysville, Maryland

Mike Wallace referred to such a patient's "biologic uselessness" at least four times. I find this frightening. Doesn't Wallace know that in Nazi Germany the identical concept led to the murder of hundreds of thousands of "useless eaters" such as the retarded, the disabled, etc.? When a person's right to life is determined by his "usefulness" under someone else's standards, we'd all better look out.

Anne Collopy
Minneapolis, Minnesota

I can't afford the available medical profession now.

If people keep trying to become millionaires because someone isn't God, the available medical care will become even more unaffordable for people like myself.

Jerry Z. Heston
Walworth, Wisconsin

You have missed a critical point in this case. The doctors, and the hospital, were part of a Health Maintenance Organization (HMO). The patient only paid two dollars for the care, or lack of care, received. The doctors did not conspire to kill this patient, but they probably did receive considerable pressure from the administrator of the HMO to hold the cost of the intensive care unit to a minimum. The victim in this case was *not* the object of a sinister plot by the doctors. He was the first recognized victim of "socialized medicine" and hospital cost containment in the United States.

Wayne Socha
Monrovia, California

I believe the root of the problem lies in the absence of consistent, nationwide guidelines for dealing with this type of problem. High technology is the blessing of this age; however, unless man can use it sensibly and communicate about the moral and ethical problems that arise, we should just go back to the practice of medicine as it was a century ago. Under those conditions, the patient would have just died from a heart attack. The wife would not have had to follow her husband's wishes; the nurse would not have had her sensibilities outraged; the DA would not have gotten involved; the neurologist who never even saw the patient would not have had an opinion; and the court who overturned the dismissal might have spent its time convicting a *true murderer.*

Karen de Pianelli
Gaithersburg, Maryland

Without passing judgment on either doctor or family, I am thankful for having a family whose faith in God helped them make the right choice when in Aug. 1976 I suffered cardiac arrest and breathing stoppage. Too, I am thankful the doctors did not after 72 hours in a coma advise my family to remove me from the life support system.

As for permanent damage I do suffer chronic kidney failure and must have hemodialysis three times a week to sustain life; I further have some memory problems. But I am thankful to see May 8, 1983.

Peter Williams, Jr.
Miami, Florida

I, too, agonized over the decision of having my husband's life support systems removed. NOT because of hospital or doctor care.

But the question will always haunt me . . .

"Was there a chance?" Only God knows for sure!

The horror of looking at the massive head injuries; his swollen head, broken bones—I KNOW that body was NOT my husband. Therefore, I found strength at the time to carry out my husband's wishes.

Mrs. Jimmie McKee
Sierra Vista, Arizona

The Catholic Church teaches that doctors must use ordinary means to preserve life, not extraordinary.

Father John Cogavin
Westford, Massachusetts

What a strange world this is: Hundreds of thousands of abortions take place annually in North America—legally. Yet two doctors are charged with murder in what seems to me to have been the normal discharge of their duty, namely, the withdrawal of life-support systems for a gentleman who, in their opinions, had little hope of recovery.

E. Bruce Ross, BA, Minister
Chatham, Ontario, Canada

What a disaster! A nurse who thought she was being a hero blowing the whistle on what she thought was inappropriate treatment. This plus two eager beaver DA's blowing everything out of proportion. These actions will bring years of agony to many people: the family of the patient, the nurse, and the two doctors. All these people have to look forward to is depositions, EBT's trials, appeals, and more trials plus millions in legal fees. Again, just the lawyer will win.

Ralph J. Argen, MD, FACP
Tonawanda, New York

To keep a body breathing and a heart beating when there is no brain function is Frankensteinian, not medicine in the great Hippocratic tradition. Hippocrates said, "It appears to me a most excellent thing for the physician to cultivate Prognosis; for by foreseeing and foretelling, in the presence of the sick, the present, the past and future . . . he will be the more readily believed to be acquainted with the circumstances of the sick; so that men will have confidence to entrust themselves to such a physician. And he will manage the cure best who has foreseen what is to happen from the present state of matters. For it is impossible to make all the sick well."

As a pathologist, I have seen resuscitative attempts of bodies with rigor mortis already present, and this was strictly for fear of the nurses, family, or ambulance personnel accusing an intern or resident of not trying everything. What happened to the capacity of the physician to declare a body dead? The Saikewicz decision in Massachusetts almost left the local judge as the only one who could call the body dead. This is a grave departure from societies' long charge to physicians to be able to tell the living from the dead.

In no way can a big suit against hospital and physicians bring back the dead, but society seems to think that health caretakers are guaranteeing good results in every case undertaken, and so wants to extract punishment when there is basically a failure of a body to meet the exigencies that stress us. This again seems to be legalistic mischief, for the lawyer is the only one to gain anything (the family's continued heartache and probable rationalization of motives will not be compensated by any financial gain they get).

Thank you for this public presentation of a basic problem—how do we make doctors and what powers do we give them? There are at present, whole schools given to the idea that Ethics and Ethicists have answers. Churches and the courts all want their say—"We are watching, doctor!" Are they watching the committee that picks students to be doctors? Are they monitoring the ethical training a medical student gets? Are they watching the greatly deteriorated interrelationship (almost to an adversary position) between medicine and nursing? If not, they are not getting to any roots at all!

The Rev. Robert W. Bain, MD
Westborough, Massachusetts

As an RN who works in a intensive care unit, I watched your story with a great deal of interest.

One of the doctors made the comment that withdrawing and withholding treatment happens frequently. That certainly has not been my experience. The patient, even if he will be impaired, *always* deserves the benefit of any reasonable doubt.

Sally Beach
Lake Worth, Florida

I should like to support the Doctors in your report, May 8, 1983. I sat by the bedside of my husband in ICU for twenty-one weeks watching him die (cell by cell) a slow, horrible, agonizing death. He was tube fed, on a respirator the entire time with absolutely no hope. This is *not* dying with dignity, and in my opinion, neither is it the will of God.

Nellia L. Miller, RN
Palatine Bridge, New York

With almost 40 years of bedside nursing behind me, I find it unconscionable that the doctors are to be tried for the murder of a patient who was declared brain dead. What is needed is a clearer definition of "death," not a murder trial. Nothing is so cruel as to commit a once dignified human being to a slow ghoulish deterioration until there are no human qualities left. We may be more concerned with ourselves and our philosophies when we prolong the dying of an already dead person.

Terry Forgione, RN
Rego Park, New York

That nurse should return to administering medications, for she lacks the education and background to be aware of the responsibilities and implications of "pulling the plugs." The doctors acted responsibly, according to guidelines in their state, and should be commended for doing the dirty work that technology has produced.

Sharon Vincent, RN
Grand Rapids, Michigan

Here I am in the December of my life hoping I could sign a legal document that would prohibit any agency from prolonging my existence with some mechanical monster!

How dreadful to think that I might be kept dying for years—fattening hospital coffers and ruining my family and their good memories of my real life.

A pox on those ridiculous young whippersnappers.

M. Clark
Central Point, Oregon

"Two Families"

Clarence Herbert had lived fifty-five years when a medical crisis brought technological intervention and a harrowing decision. But such decisions more and more often must be made at the beginning of life as well.

On December 26, 1982, "60 Minutes'" Ed Bradley talked with two couples who had given birth to babies with severe mental and physical defects, discussing the very different decisions they reached regarding their children's medical treatment.

John and Susan West live in southern California. John is a physicist at a defense plant; Susan had worked as a grade-school teacher. An attractive couple, during the interview they huddled close together on their living-room sofa while explaining in front of "60 Minutes'" cameras why they did what they did.

Their son Brian, their second child, was born with Down's syndrome, a weak heart, and a missing esophagus. The doctors told them he would need surgery at once in order to live, and the Wests tried to prevent the operation. It would be a rare operation with no guarantee of success. The hospital refused to cooperate with their wishes, threatening legal action, and the Wests decided not to fight them.

Said Susan West, "He cannot eat today, twenty months later. He's had five surgeries. He's had infections. He's had pneumonia. His heart has stopped once. So many things, painful things . . . I cannot believe that this is preferable [to death]." On camera, viewers saw a tiny infant tied down in a crib in a hospital neonatal ward, head bloodied, fluid bubbling out of a tube in his neck, surrounded by blinking TV monitors and other electronic medical equipment.

Eventually what the Wests had hoped for—that they could just "let him go"—happened, and Brian died despite the doctors' efforts.

Miles away, in Houston, Nancy and Edward Huckaby gave birth to a second child who was deformed. Edward is an architect and Nancy was an officer for a mortgage company. Their daughter Kristin was born mentally retarded and with heart and intestinal defects. The Huckabys chose to get all the medical help they could, which has included several operations that, they say, give Kristin a fifty-fifty chance of surviving beyond age two.

Said Edward Huckaby, shown lifting a smiling Kristin in the air in the family living room, "We started at zero, so every day is like the blossoming of a new flower." Nancy added, "None of our children asked to be here. It was our decision to have them. Therefore it's our responsibility to raise them the best possible way that we can."

Brian's father, John West, had said, "I view [our decision] as an act of love and genuine concern for that child. Let him go. No heroics."

Kristin's father said, "Whatever [happens], she's had an opportunity to live the fullest life possible while she was here, and that's all a parent can give a child, in our opinion."

More than 600 viewers wrote in to give theirs.

My family experienced this situation twice, once with myself and once with my youngest brother, Trevor. When I was one year old, I contracted atopic dermatitis, a severe skin rash. Due to my scratching and the extreme possibility of infection, several doctors told my parents to give me up as dead and to let them use me for some of their "experiments." My parents went to doctor after doctor seeking a treatment. During this time, I was cared for by my mother on a 24-hour basis. For almost two years, she got next to no sleep. Finally, they were able to find a doctor with a treatment that fully cured me.

My youngest brother was born with pyloric stenosis, a condition where the lower stomach valve is sealed shut, so no food can pass to the intestines for digestion. At the age of 12 days, he was starving to death. Again, my parents refused to give up on him, and he was saved by delicate surgery at the age of 18 days. Later in his life, he was diagnosed as being hyperkinetic, a controllable chemical imbalance which affects learning abilities. He was tutored, cajoled, punished, rewarded, and pushed down the right path over the years, and now serves in the United States Navy.

The person whose life is involved (the baby) does not get a say in the decision; who knows what that may destroy? Look at Stephen Hawking, the great mathematician who can hardly speak, sit up or eat, yet is rewriting some of Einstein's work.

Mark D'Gabriel
Brampton, Ontario, Canada

Exactly five years ago, my husband and I were faced with the same decision—except that for our premature twins there was no surgery necessary, only the chance to gain enough weight to sustain life.

From the start doctors in the Neo-Natal Intensive Care Unit at University Hospital in Seattle held out little hope, but they still tried—17 hours for our son and 3½ weeks for our daughter. At no time did we tell them to stop.

On reflection I think we did a disservice to Melissa when the odds against her were so very high, but she was our daughter and we so badly wanted her to live.

J. Ellyn Cook
Seattle, Washington

How much were the fees collected by the surgeon and other neonatology team doctors who prolonged the life of that Down's syndrome baby?

Henry J. Palacios, MD
McLean, Virginia

In "Two Families" you were comparing apples and oranges. Brian probably suffered every minute of his life. You were *very* tough on his parents.

Our son is 36 with multiple problems. Doctors told us to institutionalize him when he was 6 months old and later when he was 20. We didn't and are not sorry, but we do worry about what will happen to him after we die. This country can afford any dizzy new military system but damn little for good care for the millions of physically and mentally handicapped. Conservative right-to-lifers are the first to vote against such funding.

Who paid for Brian's hospital bills? And who would have paid if he had continued to live?

Nick Coniaris
Hollis, New Hampshire

I have been personally involved in such a decision making process. In addition, my professional involvement as a plastic and reconstructive surgeon in a major university center brings me in frequent contact with similar neonatal problems such as you described.

The use of such words as "kill" and "starve" is uncharacteristic of your usual reporting techniques. I think Mr. Bradley should publicly apologize to the families involved for subjecting them to the necessity of answering such questions.

James G. Hoehn, MD
Menands, New York

It was refreshing to hear a national newsperson refer to withholding medical care from an infant as exactly what it is—"killing the child."

I will know that the national news media has fully come to its senses when it refers to abortion in the same way—"killing the child."

John R. Price
Attorney at Law
Carmel, Indiana

I have my viewpoints on these issues. Regardless, you are to be commended for an excellent presentation on both side of a very delicate and real dilemma facing many parents of retarded, physically damaged, children. Your interviews were direct, harsh, and gentle on both sets of parents. It was, indeed, television reporting at its very best. Thank you for giving me the other side of my own opinion.

Robert L. DeBruyn
Manhattan, Kansas

It is sad that these parents were denied the right to let their child be released into a better experience. It seems to me that very few Christians really believe in what they preach. Our handsome, physically perfect, much loved, 19-year-old son was shipped off to Viet Nam by our government to be killed in a political war. I wonder how many of those who raise their voices in horror over the parents who let their deformed children die, even think about the government allowing the killing of our youth in their prime.

I sometimes wonder where our values are when people fight for the right of an unwanted, unborn child to be born, or a deformed child to live, but do nothing about the slaughter of our youth in unnecessary war; on the highway by drunk and drugged drivers; or just starving to death in today's economy. Why don't all the anti-abortionists fight as hard for the living?

Natalie L. Cochrane
Freeland, Washington

Perhaps the feelings of the West family, who had chosen to let their severely handicapped baby die, can further be expressed through the words of Pearl Buck. In her book, *The Child Who Never Grew,* she wrote the following passage about her own severely mentally retarded daughter.

> All the brightness of life is gone, all the pride in parenthood. There is more than pride gone, there is an actual sense of one's life being cut off in the child. The stream of the generations is stopped. Death would be far easier to bear, for death is final.

Carol A. England
Lynchburg, Virginia

The couple that preferred to let their child die without medical treatment apparently feel that life should be without inconvenience, hardship, suffering, defects, physical and mental imperfection.

The couple that felt gifted with a child that was physically imperfect view these problems as part of the struggle of life. They believe that a human being is a gift in itself.

The latter couple has the more genuine, wholesome attitude and we like their concept of love.

Dean Budde Family
Avon, Minnesota

The medical team undoubtedly made many valiant and heroic efforts to create something that in our present state of knowledge was not possible, and are to be praised.

However, I wonder whether those who "forced" the decision that resulted in that child undergoing all of those operative procedures, with the massive suffering that entailed for the child and his parents, who obviously loved him, may be guilty of *child abuse?*

Harry Shragg, MD
Los Angeles, California

Like the Wests I am a member of the Lutheran church and I think the West's decision would be a minority viewpoint. I believe I'm going to a better place after I die too, but I would pray that my IQ wouldn't be the determining factor to hasten my demise.

Mary Hussmann
Little Rock, Arkansas

As a practicing pediatric cardiologist, I have seen many children with Down's syndrome. You implied that with proper medical attention and early stimulation, many children with Down's syndrome can be "normal." In fact, a rare child with trisomy 21 may reach a "normal" IQ of 100; most are in the 70–90 range. More reasonable expectations are that a child with Down's syndrome may dress himself with help, may feed himself (but not cook), and may talk with a limited vocabulary. These are "busy" children who do not know right from wrong and must be closely supervised all of their lives. Even if healthy, they require so much attention that the normal children of the family are ignored and may develop emotional problems. Sometimes the situation leads to divorce. Guess who takes care of them when we fix them and they outlive their parents? (The siblings or the taxpayers!!)

Second, I suspect that in a few months, you will be running another program on the high cost of medical care and insurance; yet you have just advocated spending hundreds of thousands of dollars to repair the hearts and gastrointestinal tracts of children we know will still be mentally retarded. Please tell us taxpayers and insurance payers how much money we spent on those two children with Down's syndrome; money which we no longer have to spend on children with normal IQ who might have a chance to be productive to society.

It is hard to know where to draw the line in caring for children with known chromosomal abnormalities and mental retardation. Sometimes doing nothing may be right for the family and for society. Please present that side more fairly.

I will sign my name but do not want it used—we are currently doing what families wish.

A physician affiliated with a major midwestern medical center

I am the mother of a 2½-year-old girl who was born with Down's Syndrome and a serious cogenital heart defect. Without early medical intervention it is doubtful if my daughter would have lived 3 or 4 months. She had open heart surgery at 9 months and is now amazingly healthy.

There were two points that you made that I found of particular interest; first, that it is impossible to tell at birth now seriously mentally impaired a Down's Syndrome baby will be. At 2½ my daughter can drink from a cup by herself, she can feed herself using a spoon and a fork, she is beginning to put two words together, and she is starting to tell me when she has to go to the bathroom.

The second point was the fact that parents could choose to put a Down's Syndrome child or any handicapped child up for adoption. You see my daughter is adopted. At this point, I'm not even sure I am going to be able to get a second child because there are so many people wanting to adopt Down's Syndrome children. So I would encourage any birth parent who doesn't feel that they are able to assume the responsibility of raising a handicapped child to release the child for adoption.

Barbara Haviland
Grand Rapids, Michigan

Our job as physicians, once we have made the diagnosis, is to inform, advise, counsel, and support. And certainly not to lay guilt trips on caring and intelligent parents who have enough problems already. *This* in my opinion is playing God.

Harold L. Kayser, MD, PC
Westminster, Colorado

The story of the "Two Families" caused me much soul searching—with me coming to the conclusion that both families made the *right decision!*

If this be incongruity to the *n*th degree, then so be it. For I do believe that each family had excellent reasons for what they did (or tried to do). And it worries me not at all that their decisions were completely and diametrically opposite.

In a more complicated future many other diametrically opposing solutions may prove to be correct—in pairings from all 180 degrees of the compass.

Harry W. Hunt
Huddy, Kentucky

CHAPTER 9

Dear Morley

It's important, first of all, to understand that Morley Safer has a unique—some might even say perverse—sense of humor.

Once a priest wrote to Andy Rooney to correct him on a grammatical error. Rooney answered the priest, apparently disagreeing strongly. The priest, offended by Rooney's response, for some reason wrote to Morley Safer to complain. Now, Morley Safer could have ignored this letter, or let his secretary handle it, or sent it to Audience Services. But if he had done any of these things, he wouldn't be Morley Safer, and the following letter never would have been written:

> Dear Father:
>
> I received your letter and on behalf of all of us here at "60 MINUTES" I offer our most humble apology. Now do you see what we must put up with constantly? If Rooney did not have a "no-cut" contract we would have him out of here in minutes.
>
> Apart from his lapses in grammar, he is insulting, ungrateful, and churlish. Perhaps he thinks his wealth and pudgy good looks allow him to get away with murder.

I have tried everything but prayer. That being your department, Father, I urge you on behalf of us all to please pray for his soul and peace of mind.

With respect.

Morley Safer
CBS News, "60 MINUTES"

Safer sent a copy of this letter to Andy Rooney, who he knew would enjoy reading it as much as he had enjoyed writing it.

Morley Safer sees all his mail and personally answers some of it ("not much"). "I answer if the mood strikes me, or if some idiot thinks I said something I didn't. The more angry they make me the more likely I am to answer." When he does, as likely as not his reply will be a handwritten note scrawled on the viewer's letter. It saves paper, and besides, as Safer thinks, "It's wonderful to make the odd viewer really pissed with you."

It irritates him when viewers misunderstand what they've seen or heard. He recalls getting a letter accusing him of racism because he had been filmed standing in front of a tennis court near a sign that said "Whites Only." Having lived in England for years, of course, Safer knows a dress code instruction when he sees one. And it irritates him when "amateur grammarians write to us—they're almost always wrong."

Nevertheless, Safer is fascinated by the mail and often deeply moved by it. He even understands the misunderstandings: When a viewer writes in who has obviously misheard something, it reminds Safer of "being in a classroom on a hot June day, dreading the teacher will call on you and ask, 'Okay, what did I just say?' We've all been there."

He has seen "60 Minutes' " mail change considerably over the years. "In the early years there was a lot of mail from people who were shut-in, whose only friend was the TV. Now that the broadcast is so popular, the mail comes from all kinds of busy people who are involved and interested in lots of things besides watching television."

Safer wishes that the letters segment at the end of each broadcast were longer. "I wish they'd use three times as many letters. It's the only place people get a chance to have their say."

Safer has gotten his share of important letters from prominent Americans, but, typically, the only letter he has framed and hung on his wall is a part of what was originally a very long letter written on toilet paper. "After listening to CBS-TV dilettantes discussing DEATH BY WAR every year for 10 years," it says in an angry scrawl, "I'm ready to boycott you, your sponsors and your *fairy* programs called *NEWS!* SHIT is too good for you bastards!!" Safer had it framed in silver with a handsome light-blue mat.

This hangs near a framed autographed photo of a man Safer laughingly identified as "one of the great mobsters." "To Morley Safer," it says, "truly a Great newsman. And a Great guy. Sincerely, Mickey Cohen."

The important letters, the ones that touch him, he keeps filed away out of public sight. Like the letter from Katharine Hepburn, who had claimed she wouldn't watch herself being interviewed by Safer on "60 Minutes," but later wrote to thank him, noting that friends who had watched it had told her it was terrific.

He also heard from Carter Zeleznik, the father of a little boy who was killed in a Florida motel by a just-released mental patient. "Marking time as we do," Zeleznik wrote Safer, "it was eight years ago today that an important part of our world came to an end. It was especially tragic for us because it made no sense. As a direct result of your efforts we are now able to reflect upon the fact that the probability of the same sort of thing happening again may have been significantly reduced." That, says Safer, is an example of "a letter that almost makes you cry."

Safer says he has learned that "the more peaceful the passion, the more crazed the mail. If you go after pet or plant lovers, you're in trouble." [See chapter 3]. What letters does he like best? "It always pleases me if people say, when interviewed, 'You were fair.' "

CHAPTER 10

Vietnam

In autumn 1982, seven years after the fall of Saigon, America was finally ready to take a closer look at the war that had so bitterly divided us. A "National Awareness Day" was scheduled. There was to be a parade of veterans in Washington, and the Vietnam Memorial was to be dedicated at last, two years after the ground for it was consecrated.

"60 Minutes" devoted two stories to the subject of Vietnam that fall, and in reopening the wounds inspired hundreds of viewers to write down some of what they feel nowadays.

"Vietnam '82"

Seven years after the U.S. embassy in Saigon was evacuated, Mike Wallace returned to Vietnam with a group of veterans "to take a look at Vietnam today." What they found was a country in a state of economic disaster, ruled by corruption, still bearing the scars of the American presence.

Wallace located the 10-year-old girl whose image was wired around the world in 1972 in a famous photograph: naked, arms outstretched, running down a road in shock and terror after being severely burned by napalm dropped on a temple by American planes. She was a lovely 20-year-old now, studying to become a chemistry teacher. She rolled up her sleeve to show Wallace the scars from her burns.

As Wallace reported, most foreign goods available in Vietnam today, cut off as it is from American or Chinese aid, come in the mail from refugees who fled. So while there's little available in official stores, a thriving black market exists where a small box of Hershey's cocoa, for example, costs the equivalent of three months of a teacher's salary.

Wallace found a number of Vietnamese who wanted to leave the country, including a couple who were willing to say so, warily, on camera. The Soviets, said Wallace, are "not especially liked in Vietnam. They are surly. They tend to keep to themselves. They're contemptuously referred to as Americans without dollars. . . . The reality [in Vietnam] is poor and bleak. But tell that to a Vietnamese official and he will answer you, in effect, 'You thought we would be defeated in battle: We won the war. Watch and see. Yes, we have trouble, but we will win the peace, too.' "

On your program someone said, "They beat the Americans," meaning the Viet Cong. Let it be perfectly clear that the Viet Nam tragedy is that the Viet Cong did *not* "beat the Americans." The Americans were beaten by our own politicians and every lost life, every drop of blood, should be charged against our own leaders. Let history be the judge. All Americans should be ashamed that they allowed our own sons to be used so shamefully by those who refused to let our Armed Forces win.

Noreen R. Heffner
Boyertown, Pennsylvania

Rather than show the horror, suffering, damage, and terror poured out upon the South Vietnamese people, you chose on your program to show AMERICAN-caused damage from napalm, or the bombing of a hospital by AMERICAN airplanes. WERE WE THE ENEMY? You seem to have the ability to portray us as such. No wonder the Viet Nam Vet is bitter. Somewhere along the way, I heard through the "grapevine" that the Viet Cong was the enemy. Was I wrong?

Chester E. Bender, Jr.
Idabel, Oklahoma

Your story unfairly placed most of the blame for Vietnam's current troubles on the communist government, as if to justify the American intervention there. Sounds like the old imperialist rag: "If they'd only let American business come in and provide jobs, everything would be fine."

Fact is, Vietnam's plight is not very different than most other Third-World nations, except for the degree of devastation caused by their thirty-year war for independence. As you well know, the unhealthy dependence on foreign exports and the corrupt bureaucracy are not unique to countries receiving most of their foreign aid from the Soviet Union.

It is irresponsible journalism to imply (as I believe you did) that most of Vietnam's current troubles stem from the policies of the current regime. You should at least consider giving equal coverage to those nations whose economies have been crippled by an unhealthy dependence on the US—such as most of Latin America.

Would it be asking too much for CBS to do a serious study of the causes, consequences, and some possible alternative solutions to underdevelopment in the Third World that goes beyond the superficiality of this Vietnam story?

Larry Lewack
North Country People's Alliance
Burlington, Vermont

I was a student at the University of Oregon from 1964–1969. During the time I was spit on for defending Dow Chemical in their business of making napalm—a weapon used by American servicemen who were doing their precious best to come home on their feet, not in a box.

My husband was one who came home on his feet—but he was shot at, friendly fire from guns trying to stop Viet Cong from coming into their barracks one night.

Your program put everything into proper focus. Our daughter, almost 10 years old, saw on TV how our men and women in Viet Nam were killed—not by the Viet Cong or Red Chinese or Soviet guns—but by friendly fire and the American Press. I refuse to accept the guilt of Viet Nam. The press could show the sins of war and burden of guilt that should be born by the Communist nations. Instead, you showed a Vietnamese girl burned by American napalm bombs—not the Viet Cong amputating children in an American-sponsored orphanage, nor a Vietnamese child approaching a group of American soldiers with a grenade in his hand, nor the evacuation plans of the Viet Cong who have 3 days to leave an area because the US will bomb their camp at that time.

Even in retrospect you focus on American wrongs. You would still be killing our young promise of America with Friendly Fire. In other words, you really "bombed out" last night.

Sherril A. Wallace
Lebanon, Oregon

The Vietnamese people are saddled with another of those police-heavy states that can't produce anything but armed forces, more wars, and shortages. That's news? I wish the people luck, and their Government the famed salute of the upraised middle finger, which is more than they deserve.

The Korean War was felt by many here to be a poor idea. I thought it worthwhile, even when I was there. Later I had a chance to visit some of the places where I'd served, and they were 100% changed. Everything looked as if the people had worked like blazes and gotten somewhere. The barren hills were covered with green due to reforestation, and in a province noted for its bad farmland, they were setting up seaside resorts and parks. I looked at the Seoul of 1978 and wondered how Saigon would look 25 years after if we'd handled that war right. Not enough Trumans, and too many McNamaras, I guess.

John P. Conlon
Newark, Ohio

Vietnam . . . Vietnam . . . Poor Vietnam!!! They got exactly what every draft dodger and anti-war protester wanted them to have . . . PEACE.

Sharon M. Minton
Salem, Oregon

I was stationed as an Air Force Flight Surgeon Advisor in Nha Trang during the war for a year and took the opportunity to travel within Vietnam, work in the local hospitals, and make many Vietnamese friends. For the life of me, I can't understand why that government chooses poverty when they could enjoy a better life. Vietnam is a beautiful country. The South China Seacoast offers crystal coral waters and sandy beaches with stunning green mountain views. Why don't they open a tourist trade, build some hotels, and invite the world back? I know the people themselves would welcome us. And the price of Hershey's Cocoa would certainly go down.

C. M. Stanley III, MD
Macon, Georgia

Your segment on Viet Nam was somehow very fitting, after a week of National Awareness. As I watched my comrades march again, some in wheel-chairs, some limping badly, others on crutches, all waving American flags, I found myself crying like a baby as Washington and America remembered us.

This was, I hope, a beginning, and an understanding by our fellow Americans that we as young men did what we felt was right, just as those who opposed the war did what their consciences dictated. The jungles of Viet Nam have now reclaimed what our engineers once cleared away, and the hope of freedom has disappeared from hearts of the Vietnamese people.

We can no longer help the people of that country, but we can do something about our own people who now suffer the effects of the war. The Agent Orange issue still hangs over our heads, a final shroud for those of us still alive, and for our disfigured children.

I fought at Khe Sanh, RVN, with "G" Co., 2nd BN., 26th Marines. I was wounded twice, and came back with a bronze star, with Combat-V (for valor, so they say). And now at 35 my life's still in disarray. My body's full of tumors, my minds full of memories I can't forget. I hope someday the war will be over for me, but the scars are a constant reminder that war really is hell, and that death is the final reality of life. Neither of which our government is concerned about.

Thank you for your insight and for not allowing the Communists or Bobby Muller and the VA to use you. Life does go on!

Mark D. Ward
Chesterton, Indiana

"Lest We Forget"

The contest to design a memorial for those who fought in Vietnam, said Morley Safer in a segment aired October 10, 1982, was "a nationwide competition open to everyone, just like the Vietnam war itself." There were a few ground rules: The use of a flag was discouraged, because the site was so close to the Washington Monument with its circle of flags. The design should make no political statement about the war. The contestants would be known to the judges only by number.

The winner, designed by #1,026, was two black polished granite walls that would reflect the park and point to the Lincoln Memorial and the Washington Monument, with the names of the 57,692 men and women who died in Vietnam inscribed in the order in which they were lost.

#1,026 turned out to be a 21-year-old Chinese-American woman named Maya Lin, the daughter of a professor at the University of Ohio who fled China when the communists took over. Her pride at winning was soon dampened by the controversy the design inspired. Texas millionaire Ross Perot, who helped bankroll the memorial fund, pointed on camera to a model of the famous statue of the marines raising the flag at Iwo Jima, and said, "*This* is an appropriate memorial to the marines who fought in World War II. Give us an appropriate memorial for the men who fought in Vietnam."

Then-Secretary of the Interior James Watt agreed—and had the last word over the use of federal land. He demanded that a statue and a flag and an inscription be added to Maya Lin's simple monument. The memorial fund commissioned a conventional, heroic statue of three soldiers (two of them black, one white) and announced that flag and statue would stand front and center.

As Morley Safer noted, "In life, it could be argued, the fifty-seven thousand were the pawns of those fighting political battles on the home front. In death, they are still being used thus."

Safer interviewed Jan Scruggs, head of the memorial fund. "I think the difficulties we had would have been considerably less," said Scruggs, "if [the same monument] were designed by an Anglo-Saxon male."

As for Maya Lin, when Safer asked "a silly question—how Chinese are you?" she answered, "as apple pie. Born and raised in the Midwest, surrounded by non-Chinese people . . . I looked at myself as just another kid."

Many months after "60 Minutes" investigated the controversy, a compromise decision was reached: The flagpole with its 12 by 18-foot American flag stands in the entrance plaza, not "front and center" to Maya Lin's design but over 100 feet away. The statue has been changed to include one black man, one white man, and one Hispanic man. It, too, is placed about 100 feet away from the memorial. It was unveiled in November 1984, on Veterans Day.

I served a year in Vietnam as a helicopter pilot in 1970. As I became intimate with the war and the Vietnamese, I realized that we were wasting ourselves on a society mired in thousands of years of culture radically different from our own. One could sense the terror among one's comrades that they would die for naught. But most prevailed and conducted themselves with honor. We pulled together and struggled so that each might have his chance to return home. Some did not.

I had hoped that a memorial would be an image reflecting the sacrifice and bonding together a group of Americans made in a savage war halfway around the world. I was bitterly grieved when the "winning" design was announced: a black tombstone in a depression. I almost cried. It was not a memorial to our veterans; it was a black granite thorn in the heart of America reminding us not to engage in such folly again. The monument elegantly portrays the latter but fails miserably to symbolize the brotherhood of a small group of Americans who struggled and sacrificed so that each would have a chance to return home.

James J. Hutson, Jr., MD
Miami, Florida

I spent much of 1964, 1965, 1968, and 1969 in Vietnam. Far too many of my friends and neighbors have their names inscribed on the walls of that Memorial. On more than one occasion I came close to qualifying for that honor as well.

The original design for the Memorial is original, attractive, and fitting. I am not authorized to speak for the dead, but of the names listed on those walls, the ones who were my friends preferred grass and trees to tin men. I am sure that there are many of us who wish that the people who insist that another trite statue is the only fitting way to honor the Vietnam War dead would stick their statue in a musty old museum, or somewhere else where the sun never shines.

David H. Myers
Lawton, Oklahoma

A few weeks ago I paid special attention to a news photo of the proposed 3-man statue. One man, obviously white, conspicuously armed only with his mind, stood up front of two other men who were noticeably non-white, followers, armed with guns, and obviously prepared to carry out orders. Thanks to your show I see the statue differently now. It represents the politics of Viet Nam. What a reminder.

Larry Adams-Walden
Chicago, Illinois

How fitting it will be if the people with money and power win out over the woman with ideals. That is why the 57,000 died in the first place.

Joel Walz
Athens, Georgia

Americans need to (and should) take time to think about why we entered the war; why we continued the tragedy so long; why we treated the returning veterans so badly; why we were not aware of the conflicting emotions they had when they came home; and why we did not welcome them with open hearts and help them re-enter American society.

The war is over. In order not to enter another "quagmire," let us think first and maybe we won't leap next time. The memorial designed by Maya Lin with its grassy green space offers a peaceful and uncluttered environment in which to think these questions through.

Judith Mason
Brookline, Massachusetts

Ross Perot and James Watt have added the last sorry footnote to one of the sorriest chapters in American history. In our opinion they are the perfect marriage of tasteless money with tasteless power.

As for the West Pointer who started the howl—he is a perfect illustration of what happens to an individual when the arts are not a significant part of the curriculum.

Harry & Helena Semerjian
Lunenburg, Massachusetts

As a Vietnam Veteran, I feel a quite suitable monument to the sacrifice of my friends would be the help various agencies might give to Vietnam Veterans experiencing Post Vietnam Stress Syndrome to enable them to once again lead stable and productive lives. If all of the money spent on that tombstone was put into Outreach or some similar program, just maybe, one individual might surface in our generation with our experience who could provide the leadership to lead our country into a promising future.

Doug Rokke
Rantoul, Illinois

I represent the eleventh generation from five different Mayflower passengers, including their spiritual leader, Elder Brewster.

My wife's father came here from Scotland, and her mother came here from Finland. She is 100% American. So am I. No more, no less: Native American.

So is Maya Lin. THIS IS WHAT AMERICA IS ABOUT; this is what being American means. We are a nation of immigrants. We often assign ourselves a priority reflecting our family's American longevity; this is imagination.

I served a year and a half of my WW2 hitch on Saipan. Among other things I watched the wounded returning from Iwo Jima 24 hours a day for a week, unloading from hospital ships and transferring by ambulance to the island hospital up on the hill. These were heros from a valiant battle in a war critical to our existence; they and their dead buddies required a heroic monument to a heroic war, no holds barred.

Although the equally wounded and equally dead of Viet Nam were equally heroic in person, they served not in a heroic war but rather in a political travesty. If ever anyone needed a monument of "thoughtful grandeur" to reflect the idiocy of their magnificent sacrifice—and to hopefully militate against a recurrence of a major conflict staged IN DEFIANCE OF AMERICAN MAJORITY OPINION—it is the Viet Nam war dead.

To try to change the monument after everything was done "decently and in order" is an insult to our "due process" of all kinds, as well as an insult to its AMERICAN designer. But to have the heaviest heel brought down on its neck by James Watt, a self-appointed dictator who couldn't get himself elected dogcatcher, is the unkindest cut of all.

This is hardly the same government "of, by, and for the people" that I went to Saipan to fight for, or that we all moved to these shores for. Perhaps it takes a "first generation Native American" to put something like this in perspective.

Henry L. Clark
Newfield, New York

As a Vietnam veteran, I do not deserve a memorial commemorating my service to the United States of America. To be honorable in a dishonorable war is to be dishonorable.

Jackson Terlaine Schwartz
Lakewood, Colorado

It is now 1:00 A.M. on October 11, 1982 and I just finished four hours of milking cows and thinking of the Viet Nam Memorial in Washington, D.C. that was on "60 Minutes" Sunday evening. When I first saw it on the program I didn't like the Memorial, but as I was milking it was on my mind and I came to see how it represented the war. I came back from the Viet Nam War 16 years ago and put aside all memories of it, but lately people and things have made it come back into my mind. Now I've been thinking about what the memorial will represent to different people.

For those blind to the war, they can drive past and see only the Washington Monument and Lincoln Memorial and look through the Viet Nam Memorial as though it wasn't there.

Vets can see the beauty of the landscape—there was beauty in Viet Nam as well as horror. The design of the black marble viewed from a distance is narrow at both ends and wide in the middle as if a huge knife had been thrust into the United States and pulled out leaving a deep dark gaping wound. In the shining darkness are the names of those whose fate is not known.

The shining black marble represents the black silk pajamas worn by those responsible for their fate. The angle of the marble is that of a boomerang. No matter how hard you throw it, it keeps coming back as does the war. The landscape coming up over the marble shows that the Missing in Action are buried in Viet Nam never to be found. When those who went to Canada go to Washington, D.C. and visit the Washington monuments and the Lincoln Memorial, they can only look from the outside in. For if they look from the inside out, in all directions they must look at the monument representing what they ran and hid from.

Tom Packer
Blanchester, Ohio

CHAPTER 11

Taxes

"Penny Wise . . ."

This is the story, as reported by Morley Safer in spring 1983, of many a citizen's private nightmare: what would happen if the IRS got on your case?

The object of IRS attention here was the Feuer Trucking Company in Yonkers, New York. This half-century-old business almost folded in 1981 when its owner told his employees he was near a nervous breakdown. He had $2 million in debts, including $.5 million owed to the IRS.

A group of employees, many of whom had worked at Feuer all their lives, bought the company for a pittance, $25,000. For two-and-a-half years they ran the company, with yearly sales averaging $2.5 million. They repaid a lot of the IRS debt, but couldn't keep up with what the IRS wanted: a weekly payment of $18,500, with interest and penalty accumulating at the terrifying rate of 26 percent.

Though the IRS cooperated in negotiating a lower payment, all of a sudden a local agent seized the $60,000 that was in Feuer's bank account, told Feuer customers to pay their bills not to the trucking company but to the IRS, and demanded a down payment of $.4 million on the back taxes. Feuer didn't have it. The next morning a team of IRS agents, accompanied by U.S. marshals, stormed the company to seize its assets and question its owners, using what the company president called "Gestapo tactics."

But the agents discovered almost nothing worth seizing at Feuer—all the trucks were leased and even the building and loading yard were rented property. There was nothing worth selling. They took the IRS tags off the furniture and told Feuer it could use its trucks and typewriters. But they kept Feuer's $60,000 and its New York State Operating License. Without the license to operate, Feuer was effectively out of business.

Fifty men and women were now out of work and collecting unemployment, and the government was "richer" by $60,000. "It would seem that no matter how you add it up, the government is losing by closing the company down," said Safer. When he asked the IRS to comment, they said, "In this matter it would not be to our advantage to comment."

Many viewers wrote in with stories of their own to tell. It is a measure of the fear the IRS inspires that some of them refused to sign their names. But a surprising number of viewers took the IRS side.

The IRS must operate mainly on the fear of the taxpaying public. While their Gestapo tactics will horrify and disgust most of us (and I firmly agree that the public has a right to know that these things happen every single day), you gave them exactly what they wanted. They lost over $800,000 in collectible taxes, and you gave them more than a million dollars in publicity.

Considering my line of work, I'm sure you'll understand my wish to be referred to only as

An Oklahoma CPA

How does an admittedly sick company with annual sales of $2.5 million or so incur a tax liability of $500,000? The IRS taxes *net income,* and with all the deductions available to a trucking company such as investment tax credit, accelerated depreciation, loss carry forwards and carry backs, etc.—not to mention operating expenses—I don't see how even a moderately healthy trucking company of that size could generate such a large tax liability.

I'm in the trucking business myself, and the situation you described just doesn't make sense. But please don't send the IRS knocking on *my* door!

Randall S. Farrar
Mt. Vernon, Illinois

You asked the employees if they thought that they might have been treated different if Feuer were a larger truck line with representation in Washington, D.C., a very good point. It might be interesting and informative to compare the IRS treatment of Feuer with whatever arrangements were made with Richard Nixon. There was some mention of his income tax problem after his resignation. I suspect that it might be barely possible that persons or firms with "connections" might be treated more reasonably than Feuer was.

If anything is done on the above, please keep my name out of it. If Nixon still has friends and/or admirers in IRS I'd rather not have them hassling me.

Name withheld by request
Milwaukie, Oregon

Your story concerned unpaid payroll taxes. The District Directors of Internal Revenue are reasonable and have backlogs of unpaid taxes to prove it, but there are limits.

The pay of the truckers at Feuer was reduced for reported income tax and social security deductions, but the monies for these items were not remitted to the IRS. Nonetheless, the workers received individual income tax and social security credits. Some probably received income tax refunds of monies that were not paid to the government.

Forgiving interest and penalties would not save this hopeless business but would wreck the withholding system. Better some unemployment claims than stacks of payroll tax returns filed without payment.

The important lesson is that the officers of a business—even in corporate form—are personally liable as trustees for the payroll taxes. Therefore, if you cannot pay Uncle Sam, quit and cut your losses.

Otherwise, you and your family will lose everything. Bankruptcy will not save you.

Seymour Davis—retired IRS agent
Lakewood, New Jersey

Not once did you mention the fact that the Internal Revenue Service revenue officers (not agents) who were involved in the seizure action are not *permitted* to give their side of the events. Your presentation does no favor to our country's confidence in its tax structure.

Mike Amick
Plainfield, Indiana

Your story on Feuer Trucking and the IRS is even more appalling in light of the fact that on Dec. 23 Congress erased $1.5 billion in back taxes from Pacific Telephone's books.

And if that doesn't make you mad enough, GTE's California subsidiary and Southern California Gas were relieved of a tax bill that totaled $1 billion when combined.

Maybe Feuer just didn't run up a big enough bill with the IRS.

Barbara J. Sievert
New York, New York

In the words of Mr. John M. Fluke, Sr., "There is no problem, however bad, which in the hands of the Government can't be made worse."

Roger C. Bowlus
Reynoldsburg, Ohio

The IRS is just as unreasonable with small business as they are with large companies. They have no mercy for anyone who has gotten in a bind.

We have been current for the last quarter and are doing our best to keep it that way. However, in collecting the back taxes they put us on a payment schedule that will be next to impossible to keep. We were told if we were one day late they will shut us down.

We are in construction and have had one big job in the last three months. The weather has shut us down almost completely for most of the last month. It would seem that it would be better to give us a smaller payment that we can make, rather than put ten more people on the unemployment rolls.

David D. Sells
Oxnard, California

I felt your story was unjust. I began my company five years ago using withholding taxes for working capital. At the time the IRS caught up with me I was in debt to them for approximately $150,000.

The IRS sent out revenue agents to find out what could be done to solve these debts, and I found they were more willing to work with me than the bank I dealt with. Today, thanks to the IRS I have been able to employ over 1,100 people and have earned the IRS 10 times more monies than the original debt.

I am grateful to the IRS for the understanding they showed me when I had only 50 employees.

David W. Longaberger
Dresden, Ohio

You are to be congratulated for your courage in presenting subjects which could be considered dangerous to your health. But surprisingly, you missed the main point. It is not the IRS that is to blame for Feuer's predicament. The blame lies squarely on the shoulders of the members of the Senate and the House of Representatives in the Congress of the United States.

For over thirty-one years I have observed the weaving, by Congress, of a tax law so complex that the entire system of revenue collection must surely cave in in the foreseeable future. It would be an eye opener to the average citizen to begin to understand how our tax laws are made and how in the last thirty years a simple income tax revenue measure has become a Christmas tree with all kinds of ornamentation that has nothing to do with simple raising of revenue.

Dean Parkins
Seattle, Washington

Would you please consider the dampening effect that the emotional slant of this type of story has on the lessons which the public should be learning from this and every recession. Is it out of style to "save for a rainy day" and "not invest more than you can afford to lose"?

Look at the people who purchased houses twice as big as they needed with the hope that the larger one would have a greater price increase potential. Or the farmers who bid up the price of land, not for its farming value, but because they expected it to keep appreciating.

We've all heard of the "bigger sucker theory" in the stock market. Should we feel sorry for the real estate suckers or should we point out their folly to help stabilize futures markets?

One last point. The State of Michigan is in sorry shape because of unemployment. Has anyone considered that the "heaven" of a few years ago is the cause of the "hell" today? Everyone (companies, unions, institutions, and individuals) would like to blame short-term bad luck for long-term bad management.

Lawrence B. Swanson
Saginaw, Michigan

I sent this "Taxpayers' Prayer" to a local newspaper, *The Jeffersonian,* in Towson, Maryland, which recently published it. Perhaps it will be of some consolation to the truckers. I know of no other way to help them.

> Heavenly Father, we beseech You in our hour of need to look down kindly on Your humble taxpaying servants, who have given all we possess to the almighty Internal Revenue Service. Grant us that we have completed our Form 1040 correctly so no power will find fault with it. We pray to God that we have added Lines 7 through 20 accurately.

And that we have subtracted Line 30 from Line 21, so our Adjusted Gross Income is computed to their divine satisfaction.

We ask You, O Lord, to report our exemptions and bless our deductions as outlined in Schedule A (Form 1040, see pages 17 through 22 of instruction booklet), have mercy on us who failed to wisely estimate our payments during the year and must now borrow from Peter to pay Paul. Blessed are they who spent more than they earned and contributed so much to the economy.

Give us strength, Lord, so that we may dwell in a lower tax bracket forever and ever (as outlined in Publication 17, the revised 1981 Edition). Yea, though we walk through the valley of the shadow of bankruptcy (see Tax Tables, Tax Rate Schedule X, Y, Z, or, if applicable, Schedule D, or G, or Maximum Tax Form 4726), there is no one to comfort us.

Robert T. Marhenke
Baltimore, Maryland

CHAPTER 12

Dear Ed

Ed Bradley can't remember getting much mail when he was a CBS newsman in Vietnam. There was more mail from viewers, a slight trickle, when he covered the 1980 presidential campaign. The trickle increased to a steady flow when he became anchorman of "The CBS Sunday Night News." He always tried to answer his mail, and usually managed to keep up with it.

Then he joined "60 Minutes."

"I never worked anywhere where there was so much mail. It's like an avalanche here," he says. At first, he tried to continue his old habit of answering everything. "I'd come in at 8:00 in the morning to look at the mail, but pretty soon I realized the futility; answering letters here could be a full-time job. Now I go to the gym and work out at 8:00 instead."

He still looks at a sample of mail, selected by his assistant: "some good ones, some bad ones." And if something grabs him he'll answer—"maybe a criticism that's so far off I feel I must correct it, or one so especially nice I feel compelled to answer. I tend to respond either to letters that make me angry or letters that make me happy."

But Bradley has a clear sense of his own priorities about mail. "You have to be careful not to take all that criticism *and* the praise you can find in such a huge volume of mail too seriously. I think of something Laurence Olivier once told an actor about bad reviews. 'Forget it,' he said. 'But remember, the hard part is to learn to dismiss the good ones, too.' You can get an exaggerated sense of your own importance if you're not careful. The bottom line is, it still costs me 35 cents to get a cup of coffee just like everybody else."

So he's hardheaded about mail. Does he have letters from celebrities? "I don't keep those." Did Lena Horne write after his notable profile on her? "I don't think so." And what about letters from politicians? "They go in the wastebasket. You can't put much stock in that sort of thing."

Bradley does find it useful, though, to keep an eye on his mail. It's taught him a thing or two, he says, about how people watch television in general, and "60 Minutes" in particular. "When we do a factual story—a 'Who Shot John' piece, or one about an industry—we often get letters that confirm what we've said. For example, after we did a piece on the unethical practices of some executive search firms, we got lots of letters saying, 'That happened to me, too.' It tends to confirm your reporting.

"On emotional issues—like the mail that came in on 'Two Families' [see chapter 8]—the mail generally splits right down the middle. People tend to identify with their own beliefs and positions. If you ask somebody a question they view as negative, they assume it reflects your *own* position. I've learned that very few people realize that in this business we really do walk down the middle; they're always assuming you're expressing your own opinion."

The mail has also taught him that people don't always pay close attention when they watch TV. He sees misunderstandings in the letters, and thinks he knows why they happen, from television-watching in his personal life. This has made him a Sunday-at-7:00 hermit. "I try to watch '60 Minutes' at home alone. Because when I'm with family or friends, I notice that people not only react to what they see on the screen, but also interact with each other. They talk. I get very upset; I'm always saying, 'QUIET! You're going to miss what's coming!' "

Bradley, the youngest of the men known affectionately around the office as "The Four Tigers," seems to be a popular target of mash notes. There have been some rather remarkable letters, complete with invitations and phone numbers. "I used to think it was kind of funny," he admitted. "But I never see them anymore. My assistant keeps them away from me."

Bradley's office at "60 Minutes" is filled with plants of all varieties. Hardly a surface lacks a trailing vine or a split-leaf philodendron. The four correspondents sit in adjacent, identically sized offices lined up along one wall, and these offices are fully visible through their glass front walls. Seen from a distance, it looks as though Wallace, Safer and Reasoner have been suffering a long drought while Bradley has been living in a rain forest.

In the midst of this green splendor, Bradley has framed and hung only one letter. It is, however, a rather spectacular one. About four feet long, it is illustrated at the top with a drawing of a fire engine and a house on fire. Firemen are working to put out the fire. This, apparently, is how a group of third-graders in Randolph, New Jersey, conceive of a newsperson's job.

About four years ago, these children had written a letter to Ed Bradley. This was before he was on "60 Minutes," in the days when he tried to answer all his mail, and he wrote back. ("I tend to respond to kids," says Bradley, "probably because I was a teacher once.") In gratitude for his response, the children created this oversized letter, illustrated and carefully lettered in pencil. It reveals, among other things, that these children had already learned a thing or two about the news of the cold, cruel world: "We have two gerbils in our class. Pumpkin is a girl and Rascal is a boy. Two weeks ago today Pumpkin had five babies. Two were eaten by Pumpkin. We learned what the word cannibalism means. When the baby gerbils were born they looked like little pink erasers."

Bradley stays in touch with the class, and visited them last year. (They're in the eighth grade.) He has promised them that if he can work it out, he'll come to their graduation.

The mail is not, he notes, the high point of his day. "But I can't imagine *not* looking at it. You can get a letter that really makes your day. When somebody writes with a specific compliment that indicates they're on the same wavelength, they've picked up on what you're doing, then you know a moment has touched somebody's life." As for moments that have touched his own life, he has no doubt that the gerbil missive is "the greatest fan letter I ever got."

CHAPTER 13

Rape

"Depo-Provera"

Early in 1984, Ed Bradley reported on the controversial use of a drug called Depo-Provera as a treatment for sex offenders. The drug, which lowers testosterone levels, in some places is being used to "treat" men who rape.

Bradley interviewed "the San Antonio rapist," who agreed to appear on camera disguised by makeup, his name not revealed. The San Antonio rapist, naked except for a ski mask, had twice broken into the house of a young divorced woman and raped her. When he came back a third time, a group of her male neighbors was ready: They had set a trap and, after a twenty-minute struggle in the alley, they captured him. At the trial, he pleaded guilty. But the jury decided to place him on probation as long as he would take regular injections of Depo-Provera. Under that condition, he will not go to prison.

Also appearing, separately, on this dramatic segment—and also disguised by makeup—was the rapist's victim. Bradley asked her if she thought prison would have done her attacker any good. "I'm sorry, Ed," she answered. "But I'm not very much concerned with what's going to do him any good. It's my problem now. I have to live with what he's done to me. . . . I would have been very much satisfied knowing that he's off the streets, knowing that it's not just some other person's turn waiting out there for him."

Then Bradley spoke to Ray Taylor, the lawyer who defended the rapist. Taylor said the rapist's victim didn't come across very well in the trial. "She did not come across as a victim who had suffered mightily. I think she came across consistent with what she underwent, which was not a particularly brutal rape, as rapes go—a relatively mild rape."

And just in case there was any viewer who might have missed his point, Taylor repeated it: "As rapes go, this was a mild one."

You failed to present any evidence that the drug has merit in these cases. The researcher mentioned no study, no results, no controls. Only rather irresponsible theory, which, by the way, is not new. We were told of none of the drawbacks of such therapy.

Worst of all, it was suggested that rape is a crime of sex and that sex urges have a hormonal control. The former is likely not true, the latter probably not true. Rape is a crime of violence, not sex. And it is a crime against women. It is what it is because of sexism, not lust.

What will correct it is a change in culture, not Depo Provera.

Don Sloan, MD
New York, New York

Where I went to medical school in Europe sex was sex and rape was well defined as sex by force "with valid medical evidence." This is not to be confused with sex on request, under mild intimidation.

As long as this is not clarified the door is left open to legal abuses and needless suffering of many innocent men. Provera is a medical miracle. If we have the tool why not use it!

George Anghel, MD
Wilkes-Barre, Pennsylvania

Isn't it significant that we don't refer to "mild" armed robberies or "mild" burglaries? What does it say about our view of women, the most frequent victims of rape, that a rape may be considered "mild" because the victim didn't force the attacker to carry out his threats of physical harm beyond the harm caused by the act itself.

As long as rapes may be considered "mild," as long as rape is treated as a crime of passion and not of violence, as long as rape victims must suffer even more grievous injury in order to be credible and worthy of our fullest sympathy, we fail in our duty to respect and protect all members of our society and continue to perpetuate discriminatory and degrading attitudes towards women.

Judd R. Spray
Chief Assistant Prosecutor
Marquette, Michigan

Why should men be treated any different than women when it comes to hormonal imbalance? Women have been using hormonal imbalances to get out of just about everything from murder on down in the name of Pre-Menstrual Syndrome (PMS).

Why can't people accept that men are as human as women and as such just as susceptible to a hormonal imbalance.

James T. Huff
Bremerton, Washington

I wrote you 2 weeks ago to disagree with the claim that high testosterone levels cause a man to rape. Since then I did some checking and found my own testosterone level. From taking anabolic steroids, it is 9 to 10 times that of the rapist in question. Still, I have never wanted to rape anyone.

I was disappointed at your choice of mail to show. None of the letters dealt with the content or issue of the story, only angry women who go nuts when they hear the word "rape." I apologize for being so angry, it's very hard to control my anger with a testosterone level as high as mine. (Joke.)

In July I broke the world record in the Bench Press in the 165-lb. div. at the Senior National Championships in Austin, Texas. I pressed 485 lbs. I owe a lot of this to my use of anabolic steroids.

Richard A. Weil
Largo, Florida

I am a psychiatrist who was moved and troubled by your story on Depo Provera. Sometimes the instincts of the guys on the block really are better than the so-called justice of our often-confused legal system. My personal belief is that when a crime is committed, the following steps should occur:

1. apprehend the suspect,
2. determine if he or she is guilty of the act,
3. punish for the act committed,
4. help the victim,
5. help the community involved to grow stronger,
6. help the criminal.

Depo Provera decreases testosterone levels—that's good to know; convicted rapists should receive it during their jail term. Such treatment possibilities should not affect the actual sentencing process. People must be held accountable for their actions regardless of their personal, social, family, or financial problems or even brain chemistry.

The issue raised in this story by "60 Minutes" is similar to many prior stories. Did a man commit brutal murders because he was mistreated as a child? Was it his increased sex drive? His Vietnam experience? The recession? Those are all stresses, not causes of crime. The media not only records events but affects events to come. I believe it would be helpful if you took more of a stand. Interview more doctors who are aware of causes but also believe in punishment. Interview more attorneys who work for justice and not just power or money. It's time to clarify our priorities.

Dennis J. Gersten, MD
San Diego, California

I have spent the last two years educating women and men on the prevention of sexual assault and am therefore well aware of the crime and the proposed solutions. Rape is *not* a sexually motivated crime stemming from a man's uncontrollable sexual urges. Rape is a violent crime stemming from a need on the part of the rapist to overpower, humiliate, and to feel in control. Handicapped women, retarded women, young children, and elderly women are raped every day. Is this sexual motivation?

Statistics show that a high percentage of rapists are married or have active sexual partners and have had sexual intercourse within seventy-two hours of the attack. Many times the rapist has difficulty achieving an erection during the act and needs to rely on oral stimulation from the victim. Furthermore, in countries where castration was thought to be a solution, many rapists continued to rape—using a foreign object!

Depo-Provera is a useless solution to a very serious problem.

Mary A. Kociela
Whately, Massachusetts

The attorney's contention that the rape victim interviewed suffered "comparatively little" is nauseating. Rapists are horrifying enough. But it is men like that attorney, whose hatred for women is frighteningly clear despite his pretense of normalcy, who terrify me.

Ms. Rae A. Lutz
Terre Haute, Indiana

Most of us would like to believe that sex represents affection and communication. It is, however, becoming more and more apparent that a great number of men considered "normal" and "successful" are using sex, through rape, incest, and harassment in the workplace, to express something very different. In this country 1 in 4 women will be sexually assaulted by the age of 18, and 1 in 3 children suffer some form of sexual abuse by a known adult.

Can Depo-Provera really treat this problem?

Emily M. Agree
Washington, DC

For the "learned" attorney from Texas to refer to the forcible rape of a woman by his client as a "mild form of rape" is to call Hitler's Holocaust a "mild Jewish resettlement." Was he not aware that it took several grown men more than 20 minutes to subdue this rapist when his third attempt on the same victim was thwarted!

During 28 years as a judicial officer I never saw a rape victim who was not seriously affected. Rape is not the manifestation of an uncontrollable sexual urge. It is the sadistic carrying out of hostility by a person who, having given up trying to control their own life, seeks to control the lives of others.

Don Brady, CCR, CSR
Nashville, Tennessee

"A mild rape"?! Come on! Is there any possible way to "mildly" murder a person?!

Sometimes I wonder who is safer these days—my relatives in their homes in Massachusetts? Or me, serving 18–20 years in the Massachusetts Correctional System?

Name withheld by request
Massachusetts

Rape is an anti-social, violent criminal act against women. Any solely biological "cure" for rapists fails to recognize its social origins. As long as this is the case women will continue to be unrepresented in our courts and exploited in our streets. Women must not be guinea pigs for drug experiments on rapists.

William Weiss
Farmingdale, New York

Mr. Taylor has succeeded where the rhetoric of NOW has failed. He has changed me into a militant feminist by the use of just two words: "minor rape." There is no woman alive, young or old, who does not live with this Sword of Damocles over her head.

Miriam Silverman
Brooklyn, New York

I don't think that man should have drawn a prison term. He should have been sent directly to the electric chair or hanged by the neck until he was dead. Until the people in this country realize how protected the criminal is, and the cost they are to taxpayers, God help us.

Bonnie Sneed
Gail, Texas

I would like to suggest you give equal time to young men who received Depo-Provera, without consent, in the womb. In the 60's, I and many other women received this drug to avoid miscarriage. My 17-year-old now has breasts (small, but real breast tissue); has a lesser math ability; has never been able to participate in sports because of poor coordination; has been unable to run like other boys his age.

Name withheld by request
Florida

If all goes as usual the San Antonio rapist will probably develop cancer (hopefully of the testicles) and subsequently sue his attorney, the State of Texas, Johns Hopkins, and the Upjohn Company (makers of Depo-Provera). And seeing how the courts operate these days, he will probably win the case.

Larry A. Gainsburg, DVM
Philadelphia, Pennsylvania

A rapist takes away something that can never be given back. When you close your eyes it's all you see. When you want to make love to a man it's hard not to think about it, though it has nothing to do with love. Nothing will ever be the same.

Rape was my first sexual experience. When other girls talk about the first time, I try not to think about it. So please don't expect any forgiveness from the victims. They should pay for the crime they committed, but they will never pay like I do.

Name withheld by request
New York

I used to feel utter sorrow when reading of men killed by the wars instigated by men. Now I try to remember the silver lining—none of those killed will anymore be able to rape, beat, abandon, or abuse women.

Name withheld by request
New York

How many women will consider going through the painful process of prosecuting these acts now? To verbally relive the horror act by act, in a court room with the perpetrator in the same room. To face questions that humiliate, bring all the pain, the fear, the helplessness to the surface.

It has been three years since I joined the ranks of the raped, and I am still unable to shop in crowded stores, still unable to relax when people stand too close to me, and unable to get out of my car if there is any man close by. Not a day goes by I don't visualize the look on his face as I stared down the barrel of a gun pointed at my head, not a day goes by that I don't hear in my mind the sound made when he pulled back the hammer of that gun as I lay on my living room floor, belly down, hands tied behind my back with the picture in my mind of my children coming home to find my naked body.

God help the people who helped set the San Antonio rapist free, with little regard to the woman who must now live with the fear. If I had an inkling this sort of thing could happen when I was a prime witness, I doubt very much if I would or could have carried it off.

One more thought: Suppose that we can correct this biochemical imbalance. In what form of aggression will the need to overpower show its ugly face next?

Name withheld by request
Michigan

Nine years after being raped, I lived with dead-bolt locks on all the doors and windows of my house and am still afraid to be alone. I hate going out alone at night, and would never dream of jogging alone, even in broad daylight. I want to take a night-school course that would help me in my career, but fear prevents me from enrolling.

Analyze and practice new methods of treatment to your heart's content, but don't pretend to tell me or any other rape victim that there is such a thing as a "mild" rape. Should you see fit to use this letter, please do not use my name. I am not positive that my rapist is still in jail.

Name withheld by request
Minnesota

I am sitting here watching! I am 61 years old and absolutely furious!!!

I was raped 40 years ago. It was a soldier, he was married and had a child! He didn't beat me, only threatened to. I cried & begged & prayed. He didn't black my eyes or tear my clothes. But I am never in a shopping center, I am never alone, not even in a grocery store, that I'm not afraid of being watched or followed. My life has been hell—I have never forgotten or gotten over this. How dare that ignorant lawyer say that woman came through O.K. It should happen to his wife!

Name withheld by request
Oklahoma

If this drug helps prevent more rapes then thank God for it. Use it, but use it while the rapist is also being punished in prison. Rape is a brutally violent, violative crime. I know—I live with the memory of a man climbing into my bedroom window and raping me. What is so painful is the defense attorney calling the San Antonio rapes "mild." If I only had the power to let him live with the terror of being violated and the threat of being beaten. I am very ashamed that an "educated" defense attorney could be so stupid. I am especially ashamed because I, too, am an attorney.

Never quite the same
California

What the lawyer meant was that the victim, like me, was able to maintain her grasp on life. We did not buckle under despite the gross experience of the attack. Because we didn't cry enough on the witness stand or need a psychiatrist to tell us men are still okay. Because we thanked the rape crisis volunteers but never attended any therapy sessions. Because he doesn't know about those nights when every light in the house has to be on so we can go to sleep.

Name withheld by request
New York

I was most upset with the attitude of the defense attorney. I wish he could for one day live the life of a rape victim. Physical scars may heal, but the emotional scars, however well hidden, last forever, no matter now "mild" the attack.

I have no embarrassment signing my name, for as the victim I am not guilty of anything.

Julia Beifuss
Columbus, Ohio

CHAPTER 14

Nuclear Arms

In the last two years, "60 Minutes" has found several interesting ways of reporting on the ongoing, intensifying debate over nuclear arms. Some of the letters viewers wrote in response to two of those segments are presented here.

"The Bishop and the Bombs"

Amarillo, Texas, a town of 150,000 in the Texas panhandle, has for thirty years been the place where America's atomic weapons are assembled, at the super-secret Pantex plant outside of town. This is cattle and oil country, and Amarillo is "an overwhelmingly Protestant town of old-fashioned patriotic values," said Ed Bradley as he began the report that aired October 24, 1982.

Pantex employs 2400 men and women. They do their jobs quietly and without fuss—at least they did until the Catholic bishop of Amarillo, Bishop Leroy Matthiesen, made those jobs the focus of a moral issue that received national attention. "I called on those who were engaged in the production and assembling of nuclear weapons," said the Bishop on "60 Minutes," "to reflect upon the moral implications of what they're doing and to consider the possibility of transferring from weaponry work to what I call peaceful pursuits."

The bishop's activism split the town. The minister of a Methodist church who publicly agreed with the bishop was asked by his congregation to leave. The United Way voted to cut off all funds to Catholic Family Services, a nonprofit social agency under the bishop's control.

The Reverend Allen Ford, pastor of Amarillo's Southwest Baptist Church and regional Moral Majority head, was among those who spoke out against the bishop. "60 Minutes" showed him at an event he organized called "Pantex Appreciation Day." The heavyset minister, dressed in a navy blue suit, addressed the crowd: "Is it moral for men to work in a plant like Pantex? Gee, do you know what Jesus told a man one time? He said, 'Let a man that hath not a sword sell his coat and buy one.' "

He sent a number of "60 Minutes" viewers scurrying to their Bibles.

Regarding the Archbishop's statement about "What he would or would not die for": While I don't necessarily agree with his position, I would defend to *my* death his right to it, as many generations have before me. Without that willingness we may find ourselves living, excuse me, *existing* in Poland, USA.

Rick Wagner
Lawrenceville, Georgia

The Baptist minister who quoted the Bible must be writing his own Bible like Jerry Falwell and the rest of the moral majority because Jesus Christ *never* said anything about buying a sword. What Jesus Christ did say: "He who lives by the sword shall die by the sword" (MT 26:52) and "Happy are the peacemakers for they shall be called the Sons of God" (MT 5:9).

Robert B. Moorman
Ft. Lauderdale, Florida

On your "60 Minutes Show" last Sunday (October 24, 1982), in your segment on the nuclear arms manufacturing plant, you featured a Baptist Minister, who claimed that Jesus Christ had said: "If you have not a sword, go sell your cloak and buy one."

I looked through my Standard Revised Edition of the Holy Bible, and I was not able to find any such quote attributable to Jesus.

I did however, find in Mathew 26:52 a quote from Jesus of:

> "Put your sword back into its place,
> for all who take the sword,
> shall perish by the sword."

I don't know where that minister featured on your show gets his information about Jesus from, but I get my information about Jesus from the Holy Bible.

I believe that television broadcasting companies (such as CBS) are obligated under for FCC license to service in the public interest, and not to be a vehicle for broadcasting false, sacrilegious propaganda, especially not on the Christian Sabbath (Sunday).

Martin Michael Bernys
Brentwood, New York

As a follower of Jesus Christ and a student of the Bible, I was shocked by the Baptist minister's quoting Christ out of context. The scripture referred to is Luke 22:35–38, "And he [Christ] said unto them, When I sent you without purse, and scrip, and shoes, lacked ye any thing? And they said, Nothing. Then said he unto them, But now, he that hath a purse, let him take it, and likewise his scrip: and he that hath no sword, let him sell his garment, and buy one. For I say unto you, that this *that is written must yet be accomplished in me, And he was reckoned among the transgressors:* for the things concerning me have an end. And they said, Lord, behold, there are two swords. And he said unto them, It is enough."

Christ was not commanding his followers to arm themselves for defense. Rather, he was showing he only needed one or two swords to *fulfill prophesy.*

At Matthew 25:52 Christ said, "For all they that take the sword shall perish with the sword." The reason for this is expressed at

2 Corinthians 10:3,4 which reads, "For though we walk in the flesh, *we do not war after the flesh:* For the weapons of our warfare are not carnal, but mighty through God to the pulling down of strongholds."

It is my sincere hope that those misguided souls in Amarillo will turn back to their Bibles and take God's and Christ's words for guidance rather than the guidance of a man willing to twist the scriptures to his own advantage.

Stuart E. Glazier
Antioch, California

I watched your program about the "moral problems" experienced by the people working at the Pantex Corp. nuclear weapons assembly plant. I feel most people do not dispute our need for nuclear weapons, given the fact of their existence in a contentious world. However, many question whether companies and communities should become prosperous and grow by building more, and still more, such weapons. I think the following excerpt from a book by the famous physicist Freeman Dyson, who knew the American weapons scientists after the war, expressed as well the dilemma of the "Pantex families" of Amarillo:

> In February 1948 *Time* magazine published an interview with Oppy [J. Robert Oppenheimer, "father" of the atom bomb] in which appeared his famous confession, "In some sort of crude sense, which no vulgarity, no humor, no overstatement can quite extinguish, the physicists have known sin; and this is a knowledge which they cannot lose." Most of the Los Alamos people at Cornell repudiated Oppy's remark indignantly. They felt no sense of sin. They had done a difficult and necessary job to help win the war. They felt it was unfair of Oppy to weep in public over their guilt when anybody who built any kind of lethal weapons for use in war was equally guilty. I understood the anger of the Los Alamos people, but I agreed with Oppy. The sin of the physicists at Los Alamos did not lie in their having built a lethal weapon. To have built the bomb, when their country was engaged in a desperate war against Hitler's Germany, was morally justifiable. But they did not just build the bomb. They enjoyed building it. They had the best

time of their lives while building it. That, I believe, is what Oppy had in mind when he said they had sinned. And he was right."

—From *Disturbing the Universe*
by Freeman Dyson
© Harper & Row, NY 1979

Robert D. Wieting
Simi Valley, California

"The Week Before 'The Day After' "

Sunday night, November 13, 1983, was the week before ABC showed its much-ballyhooed film, "The Day After," which depicted the aftermath of a nuclear war on the town of Lawrence, Kansas.

Ed Bradley's segment that night dealt with the controversy that arose before this film even came to American living rooms. Key members of the nuclear freeze movement in the United States had somehow—"We don't know how," noted Bradley—gotten hold of advance copies of the film and were using it to promote their cause.

Some pro-defense activists, led by Moral Majority leader Jerry Falwell, were furious. Falwell said, "We plan to request equal time, under the fairness doctrine, of all the ABC affiliates that carry [the film]." He also implied a threatened boycott of companies that chose to sponsor it. He accused ABC of producing "a one-sided film . . . the number-one stroke for the nuclear freeze movement ever created in this country."

Brandon Stoddard, president of ABC Motion Pictures, denied that the film was political. He called it a "what-if" movie, where filmmakers "take a certain circumstance and say, 'Supposing such happened . . . what might follow,' in a dramatic way? We don't discuss whether deterrence is good or bad. . . . We don't even deal with that. It is a movie that says nuclear war is horrible."

It is possible to watch the "60 Minutes" segment on this controversy and conclude that it too had a point of view. But just which point of view? The first two of the letters that followed reveal that the answer lies very much in the eye of the beholder.

Does your obvious support of the right-wing element which, disregarding the First Amendment, condemns ABC's broadcast of "The Day After" represent a political statement on the part of CBS, or merely a mercenary attempt to drive viewers away from ABC next Sunday night?

Kevin J. Rust
Affton, Missouri

Well, CBS, you've finally done it. Scratch one regular viewer.

Your own pro-freeze bias was as thinly veiled as ABC's. If "The Day After" is simply a commercial venture with no political intent, then why would ABC still run it even if no sponsor supported it? A little contradictory, wouldn't you say? And why are you promoting your competition's programming? Strictly "news"? Come on, both ABC and CBS should sleep together after this cleverly orchestrated publicity stunt designed to shift American public opinion into the hands of the pro-freeze movement.

Sorry, Ed, but not all of us are blind.

Greg Becker
Eureka, California

According to the Reverend Falwell's interpretation of the "Fairness Doctrine," each Sunday morning there should be an hour program on the occult or devil worship following his own hour program, which presents only one side of the issue. After all, fair is fair.

Merilyn Zeuli
Diamond Bar, California

Your segment of ABC's shrewdly timed bit of unilateral nuclear freeze propaganda, "The Day After," put me in the unaccustomed (and uncomfortable) position of agreeing with Jerry Falwell. Please don't do that again.

Bob Pruett
Allgood, Alabama

I'm German born. I was 18 years old at the end of WWII, so believe me, I know war is hell.

I also know that Hitler prepared for war from the moment he came to power. Likewise, the Japanese were working toward conquest for many years. There would have been no tragedy of Pearl Harbor if the United States had been prepared. There would have been no overrunning of most of Europe by Hitler and his cohorts, if all of the respective countries had been ready for him. Neither of the aggressors would have been as ready and eager to attack had they known that their potential victims were as well armed and ready as they. The only way this country can be safe is by being prepared.

Nuclear freeze? Bah!! The people advocating it are dreamers. The Soviets only respect power. Anything else is interpreted as weakness.

If your neighbor has a vicious dog, you can't keep it from biting you by speaking to it softly. You build a high wall or throw it a piece of poisoned meat.

Rosemarie Kulze
Ten Sleep, Wyoming

It came as no surprise that the right wing is furious over the airing of "The Day After" by ABC television. For years the right has relied on the inability of millions of Americans to visualize the true horrors of nuclear war to campaign for a continued escalation in the arms race. Deterrence is no longer the issue, but the survivability of the human race is. Recent studies by Carl Sagan, et al, have concluded that even the explosion of 50 2-megaton warheads in the atmosphere would plunge us into a deadly "Nuclear Winter," causing worldwide death and starvation. It is time to reevaluate the entire U.S. strategic policy regarding nuclear weapons. Indeed it is likely that an all-out attack by the Soviets on the United States would cause as many deaths in the long run on both sides even if the U.S. did not fire a single shot in retaliation.

Yet with all these facts becoming increasingly clear, the right wing in this nation, led by President Reagan, insist on spending enormous amounts of our tax dollars on these tools of global suicide. The deficit brought about in large part by the arms race is a form of economic suicide as well. Young people of this nation watch as old men like Ronald Reagan mortgage their future to buy weapons we pray will NEVER be used. Many of us feel we are about to be done in by the madness of Cold Warriors a generation past.

Chet J. Graham
W. Newton, Massachusetts

Congratulations to "60 Minutes" and to CBS for putting timely reporting and human concerns above network rivalry. Of course ABC's "The Day After" is going to be a horrible experience—but not as horrible as the explosion of even one "medium" nuclear bomb. And not as terrifying, either, as the suggestion behind those veiled threats against the advertisers. The rightists seem to want to repress not only the freedom of the media but also the freedom of the economy. In saying he will run the movie anyway, ABC's president shows that he is an American who believes in what this country really means.

But where were the rightists when Hollywood made "On the Beach"? I haven't recovered from that one yet—thank God.

Katharine Lawrence
Quincy, Massachusetts

Thank you for your insightful treatment of next week's ABC movie depicting the horrors of nuclear war.

I wonder if ABC would also be willing, with or without sponsors, to produce and air a "What if . . . ?" scenario in which America, having surrendered to a militarily superior Soviet Union, comes under Communist political, social, and economic control?

Bruce D. Johnson
Central Valley, California

I live a mere eight miles from Griffiss Air Force Base in upstate New York. Griffiss was the first Air Force base in the country to receive the cruise missile. I take comfort in the fact that in the event of a nuclear war and assuming Griffiss is a primary target, I will be vaporized instantly. I do not wish to be one of the walking wounded. All the contingency plans and nuclear freezes will not save me from a mad Russian pushing a button 8,000 miles away.

Judith M. Stappenbeck
Whitesboro, New York

I cannot understand "The Day After" being termed a "political" film. Is there a party somewhere in this country which is for nuclear war? If the film neither promotes a nuclear freeze nor strengthened defenses as a means for preventing such a war (and I have read it does not) then are Mr. Falwell's objections to it based on fears it will lesson the public's taste for nuclear war? Would he censor the show if given the power? What type of show would he propose if given equal time—one which deals with a nice, clean nuclear war?

Should a network president be grilled by "60 Minutes" every time a "political" film is aired? If the American public cannot detect favoritism where it occurs shouldn't all programming (including news) be examined by a special board who would, perhaps, warn them beforehand that what they were about to see favors one point of view more than the others? Is it wrong for the networks to air what they think will attract the most viewers? Should they choose some other criteria, maybe niceness or politically centered-ness?

To sum up—I am amazed (even horrified) this film has become an issue, I object to Mr. Falwell being displayed as the Voice of the Right, and I oppose the questioning of ABC's right to air what it pleases.

Kent C. Buresh
Broomfield, Colorado

We would like to thank Mr. Falwell for his never flagging effort to control, not only our destiny, but the destiny of every living thing on this planet. We are now beginning to understand how a man so strongly opposed to a woman's right to freedom of choice can also oppose the right to life of the human race. The Nuclear Freeze Movement couldn't have found a better catalyst in any other person.

Until last night we were fence-sitters. Now, we are looking forward to learning the names of the sponsors of "The Day After." We too have ways of dealing with such impudence—by buying their products.

The Walters
Edmonds, Washington

Where are all those people who are for presenting "both sides" when movie scenes glorify war? And by the way, I am not a "liberal"—there are many conservatives who are opposed to nuclear buildup. Conserving our fragile planet goes beyond politics to essential questions of survival.

Mrs. Alice Friedman
Miami Beach, Florida

I believe ABC is to be congratulated for the guts to educate American TV viewers on the definition of "nuclear war" when it involves Kansas, not Asians or Europeans. Perhaps when we better understand exactly what nuclear war is, it will be possible to take our own political stance.

Personally, I don't think I can watch the movie. . . . I had three grandchildren this year.

Marlene Ritter
Beaumont, Texas

I wish people would get it through their thick heads that if we stop making missiles we could cause WW III. I might only be 11 years old but I think that if we stop, they may not.

Please put this on 60 Minutes, because I feel the people should hear this. And know that the little people of this world should be heard. They have some say about this. We may not be heard but at least we can say what we feel.

Yvonne Daugherty
Rogersville, Tennessee

Why make a gross show on something that won't happen. It won't happen because America isn't stupid enough to get everybody killed.

Tom Hellyer—Age 11
Chelan, Washington

P.S. I hope.

CHAPTER 15

Homosexuals in the Armed Forces

"Uncle Sam Doesn't Want You"

The Department of Defense has an official directive stating that the presence of homosexuals within the ranks affects discipline, morale, and threatens the security of our armed forces.

In a segment broadcast in January 1983 and again that June, Ed Bradley talked with several people who had run up against that directive. One was Sgt. Perry Watkins, a fourteen-year Army veteran who is testing the issue in court. Watkins, an attractive black man with a forthright manner, spoke to Bradley atop a hill on the grounds of Ft. Lewis, near Seattle. In his army fatigues and cap, he stood "at ease," hands behind his back, as he explained that when he came into the army fourteen years earlier, he had answered honestly when asked if he was a homosexual.

Watkins, Bradley reported, has had a good record in the army, receiving the highest possible scores from those evaluating his work. Nevertheless, his career is in jeopardy. Why is he fighting to stay in? "I've got fourteen years invested in the military. . . . The army has told me . . . 'You can make the military a career.' And that's exactly what I intend to do, retire from the military. I want my twenty years at least."

Also on the broadcast were three young navy women who were dismissed when the navy said it discovered they were gay. Those three, who neither admitted nor denied being homosexual, were also planning to sue.

Most of the viewers who wrote in about this issue sympathized with the gays; there was surprisingly little homophobic mail on this segment.

If it took fourteen years to find out that our black soldier was a homo, would fourteen more years really hurt?

He seems to be very smart, I hope he wins his case. By the way, I am seventy-one years old and not Gay.

George Manos
N.P. Richey, Florida

The Department of Defense directive against ANY homosexuals in the armed forces might better be addressed toward those COVERT homosexuals in high positions who are often married, with children, and have a lot more to lose by threats of exposure and a much greater chance of providing the KGB with blackmail targets.

As for morale and discipline problems, well, I do not think there are any greater problems with homosexuals than with heteros, as it still comes down to an individual's ability, or lack thereof, to command respect. God knows there are entirely too many heteros in authority who couldn't command their way out of a paper bag if it weren't for their being put into a commanding position because of outside influences or background education.

I am not generally a fan or supporter of gay rights—in fact, I have been revolted by all the gay rights marching, and what sometimes seems to be a tendency to encourage them, but in this case I am on their side and writing because of the interference with their careers. (If they were teachers of my children, I might not be so quick to advocate their rights, which shows my prejudices I guess.)

There is a saying among the truckdrivers where my husband works which says it pretty well—pardon the language, please—*"DON'T F—K WITH MY JOB,"* and that pretty well says it for homos and heteros alike. If they are good at what they do, then leave 'em alone!

Margaret Keim
Sylmar, California

What counts is character, nothing more, nothing less.

Russell W. Andresen
Orwell, Vermont

With four short years in service, I have been promoted Below-the-Zone (ahead of my contemporaries), twice received the Air Force Commendation Medal, graduated with the highest honor from a professional military school, and was selected by my Commander and a board of senior Noncommissioned Officers to represent my unit as Airman of the Quarter—not bad for a faggot, huh? If I'm a security risk or a detriment to morale, it is only by DOD definition.

I wonder if it would come as a shock to many military officials should I decide to publicly name the officers and NCOs with whom I have had casual sexual relations? However, in the interest of my professional friends, I opt to keep my private life private and allow the services to continue their outdated and prejudiced policies, so that we (the estimated 10 percent gay GIs) can continue to display our patriotism in its sincerest form—the defense of our nation.

Name withheld by request
Staff Sergeant—
U.S. Air Force

You stated that 10% of our country's population is homosexual and that probably the same percentage of our military is also homosexual.

I find it impossible to believe that the percentage can be nearly that high. Our environment must be greatly different. On your next show please identify the 10% of your program's personnel.

Lloyd Sunderland
Jerseyville, Illinois

The concept of homosexuality is repulsive. However, if a homosexual performs his administrative duties in an effective, efficient, and outstanding manner, and the heterosexual is unreliable, a drunk, and a negative morale factor, who should be recognized and promoted?

Capt. Shannon L. Trebbe
(Ret.)
Ormond Beach, Florida

As a gay in WW II, I am happy to report that we were never the victims of "witch hunts or ridicule," as in today's military establishments! We were much too busy fighting a war, without any time for "corrupting the morals of helpless heterosexuals, or playing footsie with spies." (Honestly, I wouldn't have known a spy if one had come up and bit me!) We fought, not only at home, but on the battlefields of the world. Some of us died, and some of us came home heroes. We did what we were trained to do—help win a war! Our CO's fought by our sides, and not behind peepholes drilled into the wall! And, to the hapless GI's on your segment, I say good luck in your battle with the brass, and don't back down! Stand up for your principles, and remember that there are millions of us gays out here pulling for you! So, damn the torpedos, and full speed ahead! You *shall* overcome!

Clifford Rose
Great Falls, Montana

There is one question I wish you would answer for me: Why does the government wish to make life easier in some ways for gays than non-gays? The government exempts them from military service, won't allow them to marry and pay the appropriate taxes for married couples, allows unfair business practices discriminatory to gays and therein loses the taxes those gays would pay. It would seem to me that it would be easier for Uncle Sam to give gays the same rights as non-gays if only to get from them what they get from everyone else: money.

Recently, a California commission on human rights concluded that gays could not be afforded the same benefits as non-gays because gays don't have the same kind of family structure as non-gays. They don't have children. Well, in the first place, some gays do have children; sometimes they have them from a previous marriage and in some cases they adopt. And why penalize those heterosexual couples who do not have children, because of philosophical reasons or infertility?

An old teacher of mine once said that "an organization is only as good as its weakest link." In America that weakest link is discrimination. And whether it is against blacks, whites, Latinos, women, Jews, or gays, so long as discrimination exists in this country we will never be any better than those totalitarian countries our president is so afraid of.

Susan Marie Ferrell
South Gate, California

Since the days of Alexander and Hephaestion, homosexuality has been a widespread but unacknowledged fact in all armies, prisons, monasteries, boarding schools and YMCA's.

The standard offical response has been to chastely avert the eyes, and by ignoring render it invisible.

What upsets the establishment now is *overt* homosexuality, which cannot be ignored. And there is no greater embarrassment to officialdom than being forced to admit what everyone knows already. Fact-facing has never been popular among brass-hats and bureaucrats.

Wray Wolfe
Los Angeles, California

I object to homosexuals attempting to put themselves in the same category with women and minorities. The sergeant said that years ago Blacks and women were also thought to be bad for morale. I'm a Black woman and I didn't have a choice of being either. Homosexuals do have a choice and therefore should accept the consequences of their preferences.

Jacqueline H. Roberts
Richland, Washington

Although I am not "gay," I cannot understand all the fuss about homosexuals in the military. Some of the best fighters in history, the Spartans and Thebans, for example, encouraged homosexuality; and two of the greatest generals ever, Julius Caesar and Alexander the Great, were bisexual. Certainly the United States could benefit from people of their caliber in the Pentagon.

Wayne K. Swanson
Iowa City, Iowa

Rather than a "problem," I feel the Armed Forces have a solution at hand. If our overseas contingent were to consist entirely of "gays" then the United States and other countries would not be troubled by the many thousands of pitiful, lost children discarded by the servicemen who fathered them.

Mrs. L. G. Shaw
Chelan, Washington

CHAPTER **16**

Dear Harry

"One thing you can predict about the mail," says Harry Reasoner, "is that you'll hear from people when you've scratched where they itch. For example, I did a segment on the social battle that's developed between smokers and nonsmokers recently. Lots of mail on that. It ran about five to one in favor of the rights of nonsmokers. Of course all the *intelligent* letters," he said with a smile as he reached for his pack of Pall Malls, "came from smokers."

Reasoner says he looks at all his mail. "I always have. I think if somebody writes a letter somebody ought to bother to take a look at it. Sometimes I'll answer—if something strikes me, or when the writers seem like decent people. Or maybe when I've done something wrong. Sometimes I wonder if people aren't pleased to catch us in an error. After I did that segment on Switzerland [see chapter 19], lots of people wrote in to point out that Zurich isn't the capital. It keeps you humble."

There are some categories of letters Reasoner hates: "the pointlessly nasty, or the letters that have misunderstood the point of the story. Also, letters from students, all the way from grade school up to graduate school, who write and ask you to write an essay or answer a questionnaire that would take half a day to cope with. And I don't send things to charity bazaars—I'm mostly wearing my old ties. And when people write and ask for my favorite recipe for a celebrity cookbook, I'm always tempted to answer one part vermouth and two parts gin. But I don't."

The walls of Reasoner's office are relatively bare; he is the only one of the four correspondents who does not have a single framed letter hanging. He used to, but "in a snit a few years ago I decided to put nothing on my walls but pictures of my kids [he has seven] and a Wyeth print. Everybody puts pictures of celebrities up and I just decided to get out of the competition."

However, propped beside his desk, out of sight until he pulls it out, is a large gold frame encasing three congratulatory letters he received when he left "60 Minutes" in 1970 to become co-anchor of the ABC Evening News with Howard K. Smith. It had amused him to frame them and even, for a while, to display them. On the left, on White House letterhead, is Richard Nixon's: "The news of your transfer to ABC has just come to my attention. . . . They are getting, together with exceptional professional competence, a man whose human qualities would make him a great asset to any team." In the middle is Hubert Humphrey's: "Possibly in the hurly burly of the campaign I have already sent you a note. If I did, just accept this one as an extra. ABC is mighty fortunate to get you." And on the right is Lyndon Johnson's: "CBS will be a lonelier place without you, but your friends and admirers will undoubtedly be paying more and more attention to ABC news now."

It's interesting to remember that Harry Reasoner wrote in his 1981 autobiography (*Before the Colors Fade*) that "people will do almost anything to avoid listening to a political speech, and . . . I tend to agree with them."

One politician Reasoner enjoys hearing from is Barry Goldwater. Reasoner interviewed the senator from Arizona on "60 Minutes" in 1980, and asked him whether Nixon had hurt the Republican party. "Mr. Nixon," said the outspoken senator, "hurt the Republican party and he hurt America and frankly I don't think he should ever be forgiven." After that interview was broadcast, Goldwater wrote to Reasoner, "I've had one letter from someone defending Nixon out in Missouri, telling me I was looking for a couple of fat lips, so I have to stay out of Missouri."

Goldwater wrote to Reasoner on another occasion to comment on a segment about the Navajo-Hopi land dispute; he may be the only politician who has corrected Harry Reasoner on a matter of pronunciation: "In closing, Hogan is pronounced with a long A and undoubtedly came from the Spanish word Hogar, which is the intimate of Casa. When I invite you, Harry Reasoner, to the home I live in, I invite you to my Hogar, because you are always welcome around my fire."

It seems possible that Reasoner may have decided to pare down the decorations on his office walls in an unconscious attempt to gain visual relief from the horticultural display put on by his next-door neighbor, Ed Bradley. "We call it 'The Jungle,' says Reasoner, who likes to warn visitors, 'Watch out for the tarantulas.'"

He and Bradley visit, like good next-door office neighbors. They have both noticed something they find interesting, though not especially disturbing, about letters from viewers. "A lot of people love '60 Minutes,' " notes Reasoner, "and presumably they like the four of us. But it seems they can't tell us apart. We're always getting mail addressed to each other on segments we didn't do. Ed came in one day with some mail that should have been addressed to me and said, 'How can people get us mixed up? *You* don't have a beard.' "

One thing Reasoner and Bradley have in common is that they are the only male single correspondents on "60 Minutes" (both are divorced). I asked Reasoner if he gets mash notes and he shook his head sadly. "As I said in my book, there aren't that many groupies in the news business. However, I once met a very good friend through a fan letter, and I won't say any more than that on the subject."

When *TV Guide* asked "60 Minutes" correspondents to pick their favorite segments, one of Reasoner's choices was "The Best Movie Ever Made?"—which the magazine described as "a sentimental, wry, and witty paean to the film

Casablanca." And when Harry Reasoner thinks back on all the letters he has gotten over the years, he especially remembers the mail in response to that segment. "I ended that piece by saying that most people didn't end up with the person they first saw *Casablanca* with. As I said on the air, I got engaged the night I saw it, and I didn't end up with that woman. Lots and lots of people wrote in to say, 'Oh, but I did, and we've been married over 30 years now.' It was the most touching mail I ever got."

CHAPTER 17

Pets as Therapy

"Man's Best Medicine"

Harry Reasoner, in this cuddly segment aired in spring 1983, reported on the use of animals as therapy—for hospitalized children, for the elderly, for the disabled, even for inmates in a hospital for the criminally insane. There, studies had revealed, patients allowed pets were less violent, less apt to commit suicide, and could get by on lower doses of medicine.

Other studies, Reasoner noted, have discovered that simply holding an animal can lower a person's blood pressure. "We're not at the point where doctors are going to write out a prescription, 'Rx: One dog.' But it's nice to know what we've always suspected, that in a strange and lovely way these fellows could earn their keep," said Reasoner. Then he turned to the collie he'd been petting and said, "Don't let it go to your head, Sarah."

As one viewer noted, "There are two sides to every story." Actually, there are usually even more than that.

I was interested to see that there are animals that can lower people's blood pressure. I have a cat that I guarantee would raise anybody's blood pressure and I would be willing to lend him out for anybody having a problem with low blood pressure.

Dean R. Rising, MD, FAAP
Springfield, Missouri

Thanks to "60 Minutes" for raising public consciousness regarding our value in reducing blood pressure and stress.

Last week our mistress discovered an accident on the new rug and your program saved our lives.

Peki and Bogey
Laurel, Maryland

As a proud owner of a 7-month-old Beagle, I watched your program on pets while preparing a meal of roast chicken with all the trimmings.

By the time "60 Minutes" was over, dinner was ready and on the table. I walked into the den to call my fiancé to dinner, and immediately returned to the kitchen to find my loving and affectionate puppy with the entire chicken breast in her mouth. Needless to say, she was banished from the house after a scolding. Then, to show even more how endearing she is, she proceeded to dig up the newly planted bushes and eat the insole out of a pair of shoes.

I think she was also watching your program and wanted to prove that statistics are only statistics, but dogs will always be dogs.

Lori B. Segerson
Houston, Texas

As a victim of an assault in Ohio, I was encouraged to hear about your recent program regarding pet therapy in the Lima State Reformatory for felons and rapists.

I think, however, as a matter of equal justice, the State of Ohio should provide pets to protect assault victims from further attack. Please send me one (1) German Shepard and two (2) Doberman Pinschers as soon as possible.

Tom Tomlinson
Erlanger, Kentucky

As a doctor, I think the prescription "hug a dog daily" could have more benefits than a lot of artificial methods given today. Wish I could use it, but I don't think the AMA would go for it.

That night after seeing your report, I put five more minutes on my life just by giving my dog a hug. I knew she was good for something other than hiding my shoes and scaring the mailman half to death.

By the way, I hope no one from the FDA saw your report on pet therapy. Given how they operate, they'd probably do some research studies on it showing it causes cancer.

Stephan E. Harris, MD
San Diego, California

There are two sides to every story.

I need therapy of some kind because of my husband's black Labrador Retriever. When I say either that dog goes or I do, my husband looks lovingly at his pet and says: "Well, Sam, I guess you and I will just have to go it alone around here!" Sometimes he adds, "We'll miss you."

Rosanne Washburn
Bettendorf, Iowa

Animals as therapy? Are you kidding? They helped contribute to the breakup of my marriage. My bride moved into our home with two Siamese cats, one Doberman, and a Poodle.

The cats slept anywhere they liked, including furniture, beds, and clothes closet, leaving hair all over the place; the Doberman damaged several doors and tore up the backyard, and the Poodle embarrassed me by defecating in front of Lily Pultizer's shop on Martha's Vineyard.

I'd prefer a blue-eyed blonde in front of the fireplace any day to keep me from getting lonely. Without any pets, thank you.

Name withheld by request
Marion, Maine

I am writing about your story on the animals in rest homes and jails. I think it's a good idea having them in rest homes, but a bad idea having them in jails. What if they get angry and mangle the poor animal? Another thing, I thought they were in there to be punished, not to have luxuries like animals.

I probably wouldn't like to be in jail without any friends like pets, but then again, I'd never murder anyone. Sorry about my writing but I'm only in the 6th grade, and not a very good speller or handwriter.

Theresa Morrissey
Guinda, California

Now that Harry Reasoner has done a warm, loving, gentle story on the therapeutic values of pets, let's have Mike Wallace do a hard-hitting piece on why the elderly in low income housing are rarely permitted to own pets.

Robert J. Barrett, Jr.
West Mystic, Connecticut

Do you think that the reason older people in most condo-apartment retirement communities aren't allowed to have pets is that the developers know that pets will help the retirees live longer, which makes it that much longer before the developers can sell the apartment at a higher price to someone else?

Cindy Hunt
Savoy, Illinois

If only the Federal Government in Charge of Senior Citizens Housing would view your segment. Their rules for renting a subsidized dwelling are rigid. One of these is "No Pets."

I am in my seventies and live alone with my 14-year-old dog. She is active, playful, has more pep than I have. She has been my faithful friend, protector, companion, and confidant. I can talk to her and scold her; she never holds a grudge. She gives herself in love and devotion.

To move into the Housing I would have to have her destroyed. I wouldn't do that unless she was ill and in pain. I am lucky I still have family. Some do not, and their only companion is their faithful friend, a pet. To take away their last touch with reality is heartless.

Frances H. Crudder
Tecumseh, Michigan

No doubt America's pet owners have buried you in a pile of mushy letters detailing the joy that their dogs and cats bring them.

Let it be known that there is not complete agreement on the positive value of pet ownership on every human's health and well being.

As you sat contentedly stroking your well-behaved dog on camera last Sunday evening, I was again tortured by the never-ending yelps from the neighbor's dog—stationed on permanent guard duty about 25 feet from my bedroom window.

Surely a pet can bring great comfort to its owner in times of loneliness. But what about the rest of us—when the owner leaves for the evening, with his or her dog tied up out back, with nothing to do but make noise and pass the time? Are we also adding to our life expectancy as we lay awake night after night?

Robert J. Bulka
Merrimack, New Hampshire

I am 70 years old and I have to chase my neighbor's cats (there are 10 or 12) from my house and out of my garden all the time. It makes my pressure go up, not down.

Please put this letter on your "60 Minutes" where they can see.

Mrs. Alma Starke
Jarratt, Virginia

1. Was that Harry's own dog he was petting at the end of the segment?

2. Does Mike Wallace have a dog—and if so, has anyone ever taken the dog's blood pressure?

Delores Beck
Toronto, Ontario, Canada

CHAPTER 18

Cars

Henry Ford said, "History is more or less bunk," but his internal combustion engine shaped history as surely as did any politician. Today, our feelings about cars—what they cost, what they do to and for us—reveal a complex love-hate relationship.

In the following three segments, "60 Minutes" reported on the car as killer, as a symbol of a major failure of American industry, and as a growing menace in the hands of elderly drivers—and unleashed a torrent of mail.

"Just a Couple of Beers"

A 19-year-old left a bar in Ware, Massachusetts, and got into his car. He hadn't driven a block before a patrolman pulled him over because of the squeal of his tires. The patrolman questioned him; the teenager said he'd had a couple of beers. Another policeman came along, and thought the young man seemed well-mannered and polite. They let him go.

He then drove off, and a few miles later, traveling very fast, lost control of his car and crashed head-on into the car carrying Mark and Debbie Irwin and their two young children. It was Mother's Day and the Irwins were returning home from a drive-in movie. Mark Irwin, the Irwin's twenty-month-old daughter Misty Jane, and the teenage driver were all killed in the collision.

Debbie Irwin sued the town of Ware for $1 million, and a jury awarded her $873,690 in damages—about one-sixth of the total budget of this small mill town with its high unemployment rate. Said Debbie Irwin, by then active in MADD (Mothers Against Drunk Driving) and working in a diaper factory to support herself and her son, of her lawsuit: "I think I've improved the way officials feel and think about drunk driving. There's no other way anybody would have listened."

In Massachusetts, a policeman must first arrest a person before using a "breathalyzer" to determine if he or she is drunk. And, as the patrolman in the case explained, "You can't make an arrest without probable cause." The two policemen who had stopped that teenager before his out-of-control car killed Mark and Misty Irwin and himself did not think he was drunk, though an autopsy on his body revealed a blood alcohol content twice the maximum for legal drunkenness in that state.

As Harry Reasoner noted in his broadcast that aired November 20, 1983, Debbie Irwin had brought "a lawsuit that could change police attitudes toward drinking drivers for all time." It is clear from the letters that came in that the attitudes of civilians on this subject are already undergoing profound changes.

A footnote to the letters below: The writer from Worcester, Massachusetts, was not the only viewer to notice an important detail in a scene that was on the air for only seconds.

I am a former excessive drinker who never recognized the potential danger I placed myself and others in until the aftermath of a serious alcohol-related accident in which I, driving alone, struck a parked vehicle. Miraculously, I survived 12½ hours of surgery to repair multiple fractures and internal damage. I will thank God every remaining day of my life for two facts: (1) that I was the only one hurt, and (2) that I survived and have the chance to do something about the social permissiveness of alcohol abuse which enabled me to talk my way out of over a dozen justifiable DWI (Driving While Intoxicated) situations over a twelve-year driving period. Having never been cited or punished for a crime I had committed over a thousand times, I viewed myself as a "moderate" in my circle of friends and acquaintances, many of whom had been convicted of one or more DWI violations and received not more than a "slap on the wrist."

The amount of money awarded in the Ware case relative to that town's struggling financial situation is not the issue. Nor should our sympathy for the taxpayers of Ware reduce the impact of the tragedy. If innocent taxpayers get hurt in their pocketbooks, perhaps fewer of them will die on the highways. The decision should stand up against the appeal. Changes in the standards and enforcement of drunk driving legislation are long overdue. There are millions of potential killers getting loaded in bars right this minute, and unaccountable barkeeps and cops, their accomplices, facilitating the frequency of the crime. Most of these potential killers will arise tomorrow morning with a clear conscience as I did for twelve years, and with no regrets save for a few million hangovers and exorbitant bar bills. Some will receive notification that a loved one died needlessly in an alcohol-related accident. The changes must begin soon and somewhere; why not now?

Bill Putnam
Spring, Texas

On the way home from a trip yesterday I was following a person who was driving very erratically; he criss-crossed between the median strip and the shoulder continually. The local police in New Hampshire monitor channel 9 on the CB wavelength and answered my call promptly; the Raymond police had a cruiser stop the car in front of me. I had assumed until watching your show last night that the officer would be able to give the erratic driver a breathalyzer test.

The attorney you interviewed representing the town of Ware stated that an officer must arrest a suspect prior to administering a breathalyzer test. If so, this is a legal absurdity.

Why should my family and I along with millions of other Americans have to face slaughter in our highways because the police are handcuffed by such a stupid law? If I were driving erratically, I would hope a policeman would stop me and administer any sobriety test he felt advisable, to make certain I was not endangering innocent lives.

Instead of discouraging the police from checking on questionable drivers, we should empower them to stop and question anyone driving poorly and administer sobriety tests without arresting the suspect. We want to know that the car coming towards us is being driven by a sober driver—no more drunks on our highways!

Frank Zito
Bedford, New Hampshire

I lived in Sweden for 3 months this year. People in Sweden *know* there is a stiff penalty, fine or imprisonment, if they drive after consuming more than *one* beer or wine (no "hard" alcoholic beverages). When we went out for the evening with Swedish people, one person would not drink anything and would later do all the driving. The weekend before we left, we had a "so-long" party and they hired taxis. Swedish people know the law and abide by it. A policeman there can make you blow into a balloon for any reason, if you are stopped.

Peggy Green
Garland, Texas

With approximately 70 people a day killed and countless others being injured by these nuts, we should stop playing silly legal games and crack down on the problem.

We had a problem here in Wisconsin trying to raise the drinking age from 18 all the way to 19. You probably know that a lot of the product that causes these highway deaths is made here in this state. All of this money may just possibly have influenced our lawmakers in their decision not to raise the legal age to 21.

J. L. Breedis
Green Bay, Wisconsin

Rather than wonder if the police in Ware were negligent when they stopped the drunk driver, or even if the crash was the direct result of the other driver's drinking at all, we should wonder what the Irwins could have done to protect themselves.

Mrs. Irwin talked of starting over, but will she, or anyone else who saw this story, learn anything from her experience? It is almost certain, statistically, that her husband and daughter would not have been killed had they been wearing proper safety belts.

Like every other paramedic you can ask, I have never found a dead body in a seat belt or a dead child in an approved safety seat.

The fact that Mrs. Irwin and her son were not killed may have been chance; her husband's and daughter's death were almost certainly not.

Greg Williams
Elmira, Oregon

There are no excuses for drunk driving and we all sympathize with the subject of your story for the tragic loss she endured.

It was obvious, though, from her description of the accident that nobody in her car had worn safety belts that night. When you photographed her driving away in her new car, safety belt apparently neatly tucked by her side, I had to write.

Ken Perry
Worcester, Massachusetts

Thank you very much for your piece on the Massachusetts drunk driving suit (11/20/83). The officer stated he had experience with the drunk driving problem. If that is so, surely he should have known that ANY drunk driver, when asked how many drinks he/she has had, will respond, "Just a couple of beers." Famous last words. Literally, this time. Why was this driver stopped in the first place? Wasn't it because he was driving in an unsafe manner?

Is just asking the driver how many drinks he/she has had considered the field sobriety test in Massachusetts? If so, what's the number that magically alerts the police officer that the driver is drunk? Assuming, of course, that the driver has been keeping an up-to-date tally of his/her alcohol consumption over the hours, and then is clear-headed enough (or dumb enough) to give an accurate answer to what is probably one of the most asinine questions ever asked by anyone.

I am sickened and outraged that the drunk driver problem goes on and on. Why isn't everyone? I am the mother of Michael Sulliven, killed by a member of the largest, deadliest terrorist group in America, THE DRUNK DRIVER, who kills 70 PEOPLE EACH & EVERY DAY, HERE IN AMERICA. That's 490 people killed, every single week of every single year, HERE IN AMERICA.

Pay attention, America, PLEASE!

I am a member of MADD (Mothers Against Drunk Drivers), and once again, I, along with thousands of other victims of drunk drivers, am getting ready to try to handle another holiday season without a loved one. I wouldn't even wish this pain on the woman who killed my son.

Michael's Christmas present, again this year, will be flowers lovingly scattered in the ocean where his ashes lie.

Molli McClellan
Stanton, California

In your story on drunk drivers, Nov. 20, an attorney mentioned that we should have a law prohibiting *any* consumption of alcohol before driving. As a pilot I am already prohibited from drinking any alcohol within 8 hours of flying. I wish it were 24 hours, and applied to everything from airplanes to bicycles.

James Richard Tidwell
Conroe, Texas

"... Park It in Tokyo"

"60 Minutes" took the title for this lively look at the not-so-inscrutable Japanese way of building cars from a sign outside the auto workers' union parking lot in Detroit: "300,000 laid off UAW members don't like your import. Please park it in Tokyo."

The segment, broadcast in December 1982 and again in

May 1983, was Morley Safer's, who introduced it thus: "Nissan, the maker of Datsun cars and trucks and now the third largest auto maker in the world, is building a truck plant in Smyrna, Tennessee. . . . It will be the world's most automated motor plant, even more up-to-date than these robots in Nissan's truck plant in Japan. There'll be 219 of these little uncomplaining workaholics, spinning and welding and painting—the *Star Wars* of car wars. They'll soon be turning out light pickups for rural and suburban America, already nicknamed the 'Tennessee Cadillac.' To try to ensure that an American-built pickup will be as good as a Japanese one, Nissan is spending time and a fortune to train its American workers—a year-and-a-half and $56 million to learn new skills and a new attitude."

Workers were shown during their training in Japan, doing exercises Japanese-style with Japanese workers. They waxed enthusiastic about the whole experience. Said one, "they told us to master your equipment and to love your fellow workers—it just rubbed off on everybody."

Marvin Runyon, a retired Ford executive Nissan hired to run the Smyrna plant, discussed how executives would function—eating at the plant cafeterias ("No, we don't have an executive dining room"), wearing the official Nissan uniform—for an executive, a long-sleeved white shirt with the executive's name ("Marvin") embroidered on the breast pocket and the Nissan logo on the cuff, like a bowling shirt.

Nissan claimed its wages and benefits would be as good as any UAW contract could supply and that the union need not apply. Douglas Fraser, president of the UAW, had other ideas. Interviewed in his large office, wearing a grey-flannel suit and sitting in a high-backed upholstered chair, Fraser vowed to fight: "The Japanese worker," he explained, "is obedient. He's not accustomed to exercising an independent choice or dissenting with the point of view of the foreman. And that simply is not the American way. The American worker is fiercely independent."

Hundreds of fiercely independent Americans wanted their say.

I am old enough to remember when an item imported from Japan could only be purchased in a small dime store and probably would be a tissue paper fan.

But now to *teach us* how to build autos. Have we no shame?

Paul Swanson
Shingle Springs, California

'Concern for Quality' seems to be a good slogan for Japanese auto manufacturers trying to establish a base in the USA, but judging from my experience with my Mazda GLC '82 'Concern for Quality Lemons' would be more appropriate. Moral: All that glitters is not gold and all that is Japanese is not Quality.

M. G. Rajurkar, MD
Norfolk, Virginia

I am certain that somewhere in Nissan's organization they have at least one dissatisfied employee and that person was apparently on the painting line the day my 1978 Datsun pickup truck came through. The truck began flecking paint from the hood portion the day I drove it from the showroom in January 1979.

Additionally, over these past four years, the thing appears to be moving toward self-destruction from every single seam as rust builds ever outward from these seams. Granted, in central New York cars tend to rust out faster due to the salt used to combat ice on the roads. However, I have always taken pride in my vehicles and take very good care of them; my wife's older car, which has been driven under the same conditions and given the same care, is not as close to "the final roundup" as is my truck.

If you would please be so kind as to see that Nissan receives a copy of this letter, I would be most appreciative. My complaints locally have fallen on deaf ears for years—in fact, the ears are now among the missing!

Norman I. Craner
Syracuse, New York

Maybe if Nissan builds their cars in America their dealers will be able to repair them. I have a fuel-injected 1978 Datsun 810 that only operates for short periods of time. I have been to at least two different dealers' repair shops; they still can not find the problem.

So far the only thing I find *awesome* about Nissan autos is the headaches they cause. Nissan should change their advertising slogan from "We Are Driven" to "We Are Towed."

Garry Strever
Kansas City, Missouri

You know, when I traded in my "Lemon" 1977 Mercury Monarch for a 1978 Toyota Corolla, I felt bad for the loss of my business to a Japanese auto manufacturer.

I still have the Toyota; it turned 70,000 trouble-free miles this week. And I feel that the 300,000 laid-off UAW workers "laid themselves off" in ridiculous wage demands. But an American-made car or truck with a Japanese work ethic behind it—now that's something! I get quality, and Americans get jobs. All I want to know is: How can I get an application for employment?

Norman J. Guilbert, Jr.
Greenwich, Connecticut

After driving two Japanese imports for the past seven years (with minimal trouble), in support of the U.S. auto industry we purchased a 1982 General Motors product (Oldsmobile Cutlass Ciera) for $11,000.

So far we have experienced oil pan leaks, transmission leaks, valve cover leaks, master cylinder leaks, and power-steering unit leaks—all of which the indomitable Mr. Goodwrench has been given three opportunities to repair. However, there are still leaks.

Never has any "depressed" industry deserved less sympathy than the United Auto Workers members who made the car—and the dealers who cannot or will not put forth the effort to properly repair it.

When they provide those of us who buy the product with quality workmanship, then the U.S. auto industry will regain a position of respect in the minds and hearts of the U.S. auto buyer.

Dennis E. Honzay
Sierra Madre, California

Just as every other Sunday at 7 P.M. I tuned in to Channel 2 to watch your show. This has become automatic in my house. Last night really set me to thinking about the automobiles I have owned since 1932.

1932:	a 1928 Studebaker Touring—Great
1935:	a new Ford—Well over 100,000 miles
1940:	a Plymouth which I sold when I went into the army in 1943
1948:	a Studebaker—Great
1954:	a Chevy—Great
1958:	a Kaiser—Great
1963:	a Plymouth—Great
1967:	a Plymouth—Great
1969:	a Plymouth & Valiant
1972:	a Dodge Monaco—Terrible
1974:	a Ford Torino—Terrible
1978:	a Dodge Omni—The Worst Ever

This last car was put together with spit. When I finally got rid of it in 1983 it only had 38,000 miles. It had been towed in 8 times, every part in the car had to be replaced, and when I kept writing Chrysler they sent me a check for a little over $200.00 to help pay for my misfortune with their lemon. When I tried to trade for one of their K-cars they would offer less in trade than any other company—which proved to me that they wanted no part of their own car and most certainly did not stand behind their own products.

After all those bad experiences I finally started looking at imports and my eyes were opened wide by the workmanship. I finally decided on the Datsun Stanza. I am very happy with the car. It is beautiful—rides like a dream and best of all I am not afraid to go anywhere for fear of breaking down. Where American cars were once the best they are now among the worst and they have nobody to blame but themselves.

Thank you for taking time to read this. I had to get this off my chest and give you what I think is a good reason why so many people are out of work.

Louis Stern
Far Rockaway, New York

It sounds like Nissan USA has taken a lesson from Delta Air Lines. We have always been known as the non-union airline and the money-making airline. We pride ourselves on working together toward a common goal—moving passengers in the fastest, pleasantest way—and all without the use of union officials. I hope the rest of the country can take a lesson from both of us.

Norma Johnson
Flight Attendant
Dallas, Texas

I enjoyed your segment, but feel you missed a very important point that the Japanese did not. The Tennessee worker will give you a day's work for a day's pay and take pride in the job, something learned from their kinfolk at an early age. American cars are built in the North and West where interest and support for the union far exceed that for the product.

Senter Jackson
Johnson City, Tennessee

I'm told the median annual income in the U.S. was reported as $15,500. The income of the auto worker at the time was reported as $43,000.

Pray tell me—and this also applies to Nissan—isn't the guy who earns $15,500 the same guy expected to buy the car built by the $43,000-a-year auto worker?

No wonder car sales plummet! The world's workers everywhere have priced themselves out of the market—pure and simple. Another thought—the "Automatons" at Nissan don't buy cars—ad infinitum.

Edward C. Jochens
Hodgkins, Illinois

I wonder what former President Harry S. Truman's comments would be if he lived long enough to view your show of May 1, 1983, on the Japanese building a car plant in Tennessee.

Mrs. Norman Andress
Bivalve, Maryland

If Japan is ever to have her place in the history of mankind, it would be through her repudiation of Marxism, replacing the notion of the eternal struggle between capital and labor with the harmony of the family system.

Tatsuo Hiramoto
Olmstead Falls, Ohio

Did the Japs hand pick those interviewed? Give them all an Oscar! We Yanks work to live! We don't live to work!

I'm so damned sick of the Japanese telling me I'm lazy and that we can't or don't or won't manufacture good cars and trucks. The only difference between ours and theirs is ours are big and theirs are little. I love my steel-and-leather Chrysler. I don't want a plastic-and-rubber go-cart.

Interview *me.* I go to Delco Products to work, not to be loved or patronized or condescended to.

Kathy Morris
IUE-AFL-CIO (17 yrs.)
Dayton, Ohio

What a shame the Japanese didn't win the war. We would be 40 years ahead.

Waunetta Larson
Paradise, California

Bravo!! "60 Minutes," you did it again. Just another anti-American, anti-worker, anti-labor, anti-union show. Why do you beat that same old worn-out drum? Try showing the good side of a union for once.

99% of the people you showed were young, white, and in the prime of their life. Who looks out for the young first-time employee, the partially handicapped, black, and near-retirement worker? The unions of this country, such as mine, look out for all their workers—not just the ones who can work the best for 10 or 15 years but those who can give 30–40 years to a company or trade.

Americans are so strange. They can cross oceans and continents.

They can live, eat, work, and socialize with foreigners, but they can't live next to each other in their own neighborhood without killing, burning, and raping each other.

The Japanese used bombs at Pearl Harbor, now they use everything from cars, trucks, and anything else they can to destroy our country by crumbling our economy with cheap products.

Sure they can produce cheap products by paying an average of ⅓ what the average American worker earns. How many U.S. citizens would like to work for $1, $2, or $3.00 an hour? A Datsun car might be small, but an American Datsun company can get as big as G.M. or even big enough to influence our lawmakers in Washington, D.C.

Dennis M. Dearie
Baton Rouge, Louisiana

Did anyone else notice that a so-called Christian country is sending its workers to a non-Christian country to learn about human relations?

Robert Williams
Cary, North Carolina

As a former executive of a major Japanese firm, I can state without fear of contradiction that in start-up situations the Japanese can be very brotherly—until such time as their business becomes established. At that point their strategy changes and they begin to squeeze the Americans out at all levels, replacing them with Japanese Nationals. Numerous suits have been filed with the EEOC charging discrimination against Americans and U.S. Labor Law violations.

E. J. Heil, Jr.
Atlanta, Georgia

As a union member, I find the very existence of a large organization that takes the time to care for its workers especially gratifying at a time when labor and management seem to be at loggerheads. As for Douglas Fraser, I found his choice of words suspect and calculated. The mere use of the word "indoctrination" was, I believe, offensive to the Americans who made the trip to Japan. He suggested that the average American is "fiercely independent" and that these men and women fall short of this American quality. I found his abuse of logic and his gut-level flag-waving inappropriate.

The decision of these people not to join with UAW seems, to me at least, to be a "fiercely independent" one, considering that there are 2,400 of them and 300,000 UAW's. The decision is personal, and I believe we were guaranteed the right to pursue happiness. If these workers are happy and well-treated by NISSAN, USA, then perhaps Mr. Fraser would do better by visiting Japan himself. Certainly the American auto industry could use happy and contented workers.

I would like to suggest it is Mr. Fraser who is "indoctrinated"—a natural result from reading and speaking UAW rhetoric all these years.

Eileen Land
Williamstown, New Jersey

Isn't it interesting to observe how the Japanese are so adept at putting to use what we toss out as junk? First it was scrap iron, lately our old-fashioned work ethics and moral values. It's nice to be getting some of it back.

George Nursall
San Clemente, California

Several important facts were overlooked in your segment:

- Thirty-five years ago, Japan started rebuilding its manufacturing capacity from the rubble of World War II. To this day, less than 3% of the Japanese GNP goes towards defense; the U.S. still bears the brunt of protecting Japan. The Japanese government selects "preferred" industries which receive huge injections of capital to maintain technological superiority. Compared to outdated plants in Detroit (there is only so much one can do within the confines of four walls built in the 1920's), the Japanese have an overwhelming advantage.
- Telling American workers that they will have "lifetime

time employment," thanks to the Japanese Management Style, is telling a boldface lie. For statistical purposes, only 25% of the entire work force in Japan is actually considered "full time workers." Women are never included—they are expected to stay home and raise families—nor are "yellow hats," the temporary employees who comprise the majority of the workers on the shop floor. Layoffs only affect the latter two groups, while the elite 25% stay on; this is how official statistics can boast low rates of unemployment.

There is no doubt that Japan can produce a higher-quality product, thanks to active government intervention in key sectors. The economic devastation of less-favored industries—such as ship building—rivals any U.S. auto or steel plant in the depths of the recession. In a homogeneous society such as Japan, it is possible to use company songs and group exercises to help boost productivity. However, nothing helps more than to have a friendly governmental attitude aiding the business sector.

Ultimately, the greatest Japanese export is not a car or a transistor radio—it is the myth of the superior Japanese Management Style.

Peter H. Stark
New York, New York

Yes we taught Japan how to build cars, but now it's time for Japan to teach us. So for God's sake let's open our minds and learn.

R. Stark
Saginaw, Michigan

The country's economic ills are caused not so much by Reaganomics as they are by the unions. Three major industries, steel, automobiles, and housing, are being unionized to death. Union-scale pay is so astronomical that it has no connection to economic reality at all, or to normal American wages.

For decades it has always been, "Do you know what *those* guys get?" Those of us who have had to study and bust our butts for years to get to professional competence are intensely resentful of flunkies who want to be overnight plutocrats by the use of force or even tyranny. The overpaid who strike for even higher wages should be sentenced to starting their own business with their own capital, selling over the counter in times like these, and just see what wages *they* can pay.

I am disgusted that we had to bail out Chrysler, only to find all we were doing was indulging overpaid unionists in the life style they had become accustomed to.

Boris Holland
Tarzana, California

Japan pulled a sneak attack on our country. They took a very heavy toll.

Ironically, Japan may have won the war. They have retooled from the ground up and have invaded our country in all phases of tech science. However, they are doing a very fair thing in training our auto workers and making a heavy investment in our country. It appears we will all win this time.

Donald F. Piper
Quincy, Washington

I wonder when you and others in your field will journey to Japan without cameras and fanfare to discover that under the Japanese industrial veneer of love of employees lies a fabric of feudal servitude and fear of management that permeates the Orient.

Of course Detroit was asleep—both management and the UAW are totally aware of this. But to suggest that the Japanese system is a cure-all is not only naive but very unfair to our country.

Keep an eye on Detroit, fellows; the Rip Van Winkles of Grosse Pointe and Birmingham are awake and moving, finally.

B. Hanoian
Selma, California

"Too Old to Drive?"

Harry Reasoner opened this segment by noting that, in addition to drivers who drink, take drugs, or speed in the name of adolescence, "another group—not drinkers, not drug users, not high on auto power—is being labeled as frequently dangerous to their fellow users of the roads: America's elderly, for whom a car and a license to drive it may be one of the dwindling privileges of freedom." The broadcast aired on November 6, 1983.

Some of the scenes, filmed in Florida, were shocking. Through the merciless lens of the camera, viewers watched people who could barely see or hear getting their licenses renewed. But most shocking of all was 81-year-old Gertrude Karmiol at the scene of an accident: white-haired, well-dressed, clutching her purse, she was being helped into a patrol car. Gertrude Karmiol, over a period of 39 months, had hit eleven people in three separate accidents, and killed three of them. When she finally got to court, she was given the maximum penalty: a $500 fine and suspension of the valid driving license she had been issued.

Said the judge at her trial, "In the state of Florida, you get your driver's license at age 16, you keep it till the day you die. . . . We have a written, twenty-question examination about street signs. I don't care if you're 90 years old and you can barely walk—if you can pass that written exam and an eyesight exam which they administer through a machine, you get your ticket to drive."

The judge called for a reexamination period nationwide, perhaps every four years. The letters that flooded in after this broadcast made it clear that the problem is not confined to Florida.

I am angry! Very angry!

I am a 65-year-old retired teacher and farmwife. When I was teaching I made a daily round trip of 25 miles in all kinds of weather. I have a clean record—no arrests, no accidents, but I am reasonably certain I could not pass a behind-the-wheel test. I would be so paralyzed by the threat of losing my driver's license that I doubt if I could function. I don't believe you realize what a driver's license means to those who live in the country; it's our lifeline!

What I resent most about the broadcast is the completely unsympathetic attitude. You will all, barring an early death, be elderly. Time has a way of passing quickly and suddenly there you are! After your broadcast, I feel as if we have a caste sytem—and I, since I'm 65, am an untouchable!

Mrs. Marion Foster
Harvey, Arkansas

After driving behind several elderly drivers this week, I came up with a bumper sticker I'd like to have made. There are only three things wrong with it: (1) The words wouldn't fit on a bumper sticker. (2) The drivers concerned wouldn't be able to see them. (3) It wouldn't do any good if they could.

You know you are going too slow
When the other cars are all
Whizzing by—and *they* are only doing 30.

Since I am 65 and they bother *me,* no wonder the young drivers nearly go bonkers.

Mrs. Halomae Shaw
Tulsa, Oklahoma

As a gerontologist, I must respond to the segment on older drivers. You spoke of all people sixty-five and over—an "age group" which now spans forty-plus years—as though they are a homogeneous population. That is comparable to calling the driving habits of all those between 16 and 56 those of "young" drivers.

While there are certainly some older people who are not safe drivers, this is due to specific impairments—not to advanced age itself. Your report ignored the many possible causes of poor driving—effects of drugs and alcohol, poor vision, impaired reflexes and so forth—and focused instead on chronological age.

Please, avoid ageism along with all the -isms you've been fighting so long.

Elaine Frank
Washington, D.C.

I would love to see you take issue with car manufacturers whose "new and improved" designs can easily promote accidents and confusion:

1. The four-operation gadget on the left-hand side of the steering post—it's a windshield washer, wiper of various speeds, light dimmer, and directional signal. No matter how adept you may be, time after time you will be inadvertently putting on a turn signal when you meant to dim your lights, or vice versa. Wonder why the guy in back of you honks or the guy coming at you deliberately puts on his bright lights?
2. How about putting the automatic shift back on the right-hand side of the steering post so you don't have to look down?
3. What demented brain came up with the hooded instrument panel? It's hard as hell to read. Some even are tilted downward (my 1983 Omni for instance).
4. With regard to item one, it would be nice to have the dimmer switch back where it used to be—on the floor near your left foot.

Grace O. McNulty
New Milford, Connecticut

Certainly some people at 90 are as bright, alert, and quick as they were at half that age, but they are the stalwart few. Time and the demographics are moving a larger and larger proportion of our population into the "Over 70" range. Yet the Department of Motor Vehicles continues its policy of automatic renewal of licenses for drivers in their 70's, 80's, and beyond, virtually guaranteeing them the right to drive in perpetuity.

I hope that some benevolent law will take away my right to drive when tests show that age has so diminished my reflexes, judgment, or skills that I should no longer be entrusted with the privilege to drive as potentially lethal a weapon as a car.

Perhaps I'm responding more strongly than some viewers because early last Spring my wife's car was hit by that of an 89-year-old driver who wanted both sides of the road and mistook her accelerator for her brakes. Hopefully my wife will be out of the hospital by

Thanksgiving, a little more than eight months since the accident. But out with a wheelchair, a walker, nurses pretty much around the clock, and a radically different lifestyle.

Edward S. Funsten, Jr.
Essex, Connecticut

As an eighteen-year-old college student, I was appalled by the attitude toward the elderly exhibited in last night's segment on senior citizens who drive. Perhaps some, indeed many, elderly drivers are incompetent. However, I shudder at the amount and extent of generalizations I heard. "They" are incompetent, "they stop on a dime," "they can't see," "they can't hear" . . .

If I am learning one thing from my liberal education, it is that "they" do not exist. Just as I resent being referred to as a "reckless college student," so too do I pity my middle-aged elders who dispel a generation of Americans in one broadly based statement.

And I am told that one learns respect . . . .

Dawn Skorczewski
Chestnut Hill, Massachusetts

I am 67 years of age and have recently moderated and modified my driving "habits." On trips east to Cape Hatteras, NC every summer I now drive only 8 hours per day, about 400 miles. I do not do any late night driving. Thus I find my bonus in that I can spend more time camping en route.

I spend the Labor Day holidays "buried in the sand" at the beach, allowing death to take a "holiday" also.

Before they were standard equipment on cars I adapted an airplane seat harness to my car.

I have a question:

Were Harry Reasoner and his driver, towards the end of the segment, wearing seat belts?

Jervas W. Baldwin
Des Moines, Iowa

[Answer: Their laps were not visible on camera, so it was impossible to tell if they had belts across them. However, neither man was wearing a shoulder-harness belt of the kind standard in late-model cars like the one they were riding in.]

For the last eight years I have lived in an area where approximately 50% of the population is retired. 10% are over 75 years old. Let me repeat for emphasis: over 75 years old. At least once a week I escape an accident by the skin of my teeth. I've learned to fall back a hundred yards when a senior citizen indicates a left turn because 30% of the time he turns right. I've learned to yield at any and all crossings, or face the consequences. I ride my bike and walk my dog in a state of combat alertness because drivers often do not see past the steering wheel.

The destructive trail of incompetent older drivers can be found not in the files of those over 65, but in *our* files as we try desperately to avoid collisions or attempt to pass cars going 10 mph in a 40 mph zone.

The greatest irony is that the older the driver, the bigger the car, "because they're safer in accidents." Those of us with infants in car seats and 50 years to live, who drive small compact cars because we can't afford big ones, are easily demolished by these tanks. A drunk driver in similar situations would be nailed to the wall.

Georgia Xydes (Age 33)
Kerrville, Texas

I would guess that for every accident an elderly person is in they cause five more. Either by making people pass them because they're going 45 mph on the freeway, by their hesitation at turning left, by stopping abruptly for no apparent reason, or just a general mystification they have with the road and road signs. Their right to be mobile doesn't give them the right to make others immobile.

David Chiola
Detroit, Michigan

As a practicing Optometrist, I all too often see elderly and some not-so-elderly patients who should never be allowed on the road. Some of these people do have correctable vision, but many many others do not, and yet I have no power to get these people off the road. In effect, I have become an accomplice to people committing a crime with a "deadly weapon." In Connecticut one can obtain a driver's license at 16 and drive till 116 and never be tested visually or any other way again.

In an age where we seem to be concerned about making better and safer cars—better bumpers, air bags, and safety belts—we fail to consider the auto's most important safety factor: the driver himself.

Dr. Joseph Segal
South Windsor, Connecticut

Naturally your segment was taped in Florida, where the nation warehouses its elderly. You missed one good point, however, which would be laughable were it not so frightening.

Some few years ago here the state Highway Patrol asked the state Welfare Dept. for permission to cross-check the list of licensed drivers against the list of persons receiving Aid to the Dependent Blind. Welfare, however, came down with a fit of civil rights fever, and refused permission. Your comment, please.

John Amsden (69 years old)
Bradenton, Florida

I wish the media would quit using the following: "Joe Bloke, 78, hit the truck, demolishing his car. It is believed HE BECAME CONFUSED AND HIT THE GAS PEDAL INSTEAD OF THE BRAKE." And of course you've heard of: "Joe Bloke, 17, LOST CONTROL ON A CURVE." And, finally, worst of all: "Joe Bloke, 42, HIT THE TRUCK WHEN HIS BRAKES FAILED." The young are always "out of control," the middle-aged "never make mistakes," and the old people are described as "confused." Not all of us are, anyway.

M. Jean Henderson
Charleston, Illinois

In my six years as an over-the-road truck driver, I've had the opportunity to observe most types of drivers. From fathers who are late for work to mothers trying to control rowdy children while maneuvering through traffic. From teens who think they were born to be wild, to the businessman who has to read his paper while travelling on the interstate. I've been scared out of my wits by drunk drivers as well as run off the road by one who fell asleep at the wheel.

But none of these frighten me like the elderly person whose reaction time could be measured in minutes, and who is oblivious to the traffic and pedestrians around them. I believe such drivers believe they are doing no harm. However, I also believe the problem is just as prevalent and just as deadly as the alcoholic behind the wheel.

It seems there's a relatively easy solution, but a solution that may cost a few votes that no one seems to have the courage to gamble with. Instead, they are gambling with our lives. I believe that those who suggest mandatory annual driving exams for those over 60 have the best idea. I hope the powers-that-be will someday introduce legislation that will protect the majority in this regard, as they are finally beginning to do for our alcohol problem.

Michael Gadzinski
Grand Rapids, Michigan

The older drivers, especially those in their 70's and 80's who have repeated accidents, may be the tragic victims of Alzheimer's Disease, a disease that is the fourth largest cause of death. Alzheimer's victims often continue driving unless someone stops them. They are a danger to themselves and others because they become confused and disoriented in traffic and their judgment is poor. Family members are often reluctant to take away the driver's license of these people, since driving affords them continued independence.

Anyone who cannot pass a Mental Status Questionnaire should not be allowed to continue to drive. The MSQ is one diagnostic tool used by gerontologists and geriatricians to test mental competence.

Alzheimer's Disease is an organic mental disorder, not a physical disorder, although in the later stages there is physical deterioration as well. It is a disease that dims the bright minds of young (presenile dementia) and old (senile dementia) alike.

Lenore S. Powell, Ed.D.
Gerontologist, Psychoanalyst
Author, *Alzheimer's Disease: A Guide for Families*
New York, New York

We have all seen automobiles labeled "STUDENT DRIVER," with a sign on top and on the trunk of the car. Why not require individuals over a certain age to have a sign "SENIOR DRIVER" on top and on the trunk of the car before their licenses are renewed? Most drivers would keep a safe distance away.

Cecile A. Bordlemay
Naples, Florida

As a 75-year-old driver and retired naval aviator, I can suggest from aviator experience a simple test to spot and eliminate us oldsters no longer fit to drive.

Need only be as simple as a flashing light, plus device to measure time one takes to actuate something like a hand lever or foot pedal. People like flight surgeons and psychologists have adequate data to determine the cutting line on physical reaction time for denying license. Device could also set a gray area for restriction to daytime driving, and spot the occasional younger driver whose reaction time has been impaired by illness or drug abuse.

Beauty of it is that it raises no question of intelligence. Nor does it subject testing officer to the inevitable verbal abuse were he forced to make a judgment call on ability. My reaction time is above average for my age group, but the time will come when this once young fighter pilot should no longer be allowed on the highway.

H. K. Edwards
Arlington, Virginia

The American Association of Retired Persons with a nationwide membership of 15 million, long ago recognized problems of drivers 55 years of age and over.

AARP created 55 ALIVE/MATURE DRIVING, an eight-hour classroom refresher course which deals with specific problems of older drivers. The course is available nationwide, and more than 200,000 older drivers have completed the curriculum in the first three years since it first became available.

I thought you would like to know, from a person who took the course after having an accident, that it can help the older driver remain on the road longer and be a safer driver.

Francis M. Rogers
State Coordinator
Denton, Maryland

When it costs you $15 for a taxi to and from the supermarket, when a simple half-hour errand requires three hours of waiting for bus connections, when you can't leave the house after 6 P.M. because the buses have stopped running, and when taking the subway may mean being mugged, you, too, might cross your fingers and climb behind the wheel.

Sharron Cohen
Gloucester, Massachusetts

In regards to the accident-prone oldsters: They need low-cost, scheduled, door-to-door van-bus taxis, not high-priced single fare taxis. It's always cheaper by the dozen.

Gil Hoag
Seattle, Washington

West Germany might have one answer: upon the voluntary surrender of a valid license, at age 65, that government issues a free pass on all public transportation for life. It includes all buses, trolleys, and railroads.

James May
Lake Worth, Florida

As a senior who discontinued driving a few years ago, I must ask: Why must the older generation be placed in the humiliating position of having to beg for a service that should rightfully be provided them?

Before I moved down here a couple of months ago, I was very happy and content because in Danbury, Conn., the Red Cross has a wonderful service. If you call them a day in advance and tell them where you want to go, they send a minibus to your door and, for 25 cents, they take you to your destination and also bring you home for another 25 cents. You don't have to BEG anyone to do you a favor.

But here in Greer, South Carolina, there is *no organization* that supplies transportation. As a senior I would like to participate in the activities offered at the senior center, or go to a concert, or take a class at one of the colleges, but I am deprived of these privileges.

My doctor advised me to continue swimming to stabilize my physical condition, but I cannot get to the pool. And, unless I do continue this water therapy as advised, my condition will deteriorate to the point where I'll end up as a bedridden old lady taking up a bed in an institution as a ward of the state, which will add more of a burden to the taxpayers of South Carolina. Is this cheaper than providing transportation?

Just to give you an idea of the many sources I have contacted so far:

1. Mary McCann, Coordinator of Aging Program, advised me that the Greenville Transit Authority provides a transportation-on-demand system for a minimal charge of $3.00 one way. How can seniors on Social Security afford $6.00 for transportation?
2. I wrote to Gov. Richard Riley and asked him who supplies transportation service for senior citizens who no longer drive. He turned my letter over to
3. Ms. Shuptrine, who advised me to contact Mr. Bill Boyd, Director, Greenville Senior Action, 402 E. McBee St., Greenville, S.C. Ms. Shuptrine advised me that programs for senior citizens are provided through the Greenville Senior Action, as "OLDER AMERICANS ACT" funds, as well as other federal and state funds, are utilized by that agency and transportation is provided.
4. I contacted Mr. Boyd who advised me there is no transportation service and suggested I contact
5. the YMCA both in Greer and in Greenville, which I did, but they have no transportation service, either. He also advised me to contact the
6. Greenville Transit Authority, which I did, and they too have no service yet in this area—perhaps by next year they may. He also advised me to contact the

7. United Ministry in Greenville, which I did. They service patients who must have dialysis treatments or doctor appointments.
8. On my own I also contacted another agency I had heard about called SHARE, but their transportation services are for Medicaid people only.
9. I also contacted the Red Cross, and they do not offer transportation here in South Carolina.

As you can see I've tried! Do you wonder why older people keep on driving? If they don't, then they become prisoners in their own homes!

As I pointed out in my letter to Gov. Riley, there is an OLDER AMERICANS ACT (OAA) which provides funds for supportive services under Title III—one of them is transportation. What is happening to these funds?

Mrs. Sophie Freudenthal
Greer, South Carolina

One solution to the dilemma of unsafe seniors on the road might be to require them to abandon their right to drive lethal multi-ton cars in favor of a license limited to driving a modified golf cart or equivalent. The seniors stay mobile and the people in the street are relatively safe. Besides, golf carts are easier to park.

Christa McReynolds
La Jolla, California

I am 16 years old and live in Florida. I resent what you said about teenage drivers, about us being car happy or having a lead foot, whatever you said.

Second of all, I would like to say that I don't believe that a person who is over the age of 65, blind, or like that elderly woman, having 3 accidents and nothing being done about it, can be able to drive.

I think a law should be passed to make everyone who reaches the age of 40 take a reexamination every 2 years.

Mike Davis
New Port Richey, Florida

CHAPTER 19

Places

If you really want to get people riled up, take a critical look at their hometown. "60 Minutes" has no compunctions about doing just that.

In this chapter, irate residents and amused outsiders respond to visits by a "60 Minutes" crew to four places: San Francisco; Camden, New Jersey; Zurich, Switzerland; and Polo, Illinois.

The last of these segments was not so much about a place as about a state of mind. The story of what happened one wintry day in the small farming community of Polo, and the dramatic response of viewers, constitute a kind of nationwide electronic morality play. During the period covered by this book only one other segment (on the National Council of Churches) provoked more mail than "Small Town, U.S.A.," a story of Polo, Illinois.

San Francisco

"The Different Drummer"

Late in April 1983, San Francisco was preparing to do something pretty unusual in American municipal politics: vote to recall (i.e., throw out of office) a basically popular mayor, Dianne Feinstein. A week before the vote, Morley

Safer reported on the political situation in San Francisco, and how such a thing could happen.

"San Francisco is a unique democracy," said Safer in his introduction. "Every man, woman, every practitioner of every sexual preference, is king or queen. Just to get a taste of San Francisco politics, [take a look at] Sister Boom Boom, [who got] 23,000 votes when he ran for Board of Supervisors."

"Sister Boom Boom" appeared before the camera in whiteface, bright red lipstick, and two-inch-long false eyelashes. He wore a nun's habit, complete with wimple, whose hemline ended at mid-thigh to show off legs adorned in white panty hose and plain black pumps.

Safer then unraveled the complicated story of the recall movement: A tiny band of anti–gun-control activists had started it. When Mayor Feinstein then vetoed some legislation popular in San Francisco's large homosexual community, the original tiny band suddenly could gather the thousands of signatures needed to demand a recall vote.

Through Safer's narration, the camera showed scenes of San Francisco that would never make it to a Chamber of Commerce travelogue: men holding hands and caressing each other on the street, a sign in a real estate window saying "Gay Hotel—Victorian Charm with European Atmosphere," X-rated movie theatres, men with shaved heads, and women sporting buttons that said "Support Your Lesbian and Gay Press."

In between these shots there were quiet scenes in which a very relaxed, amused-looking Morley Safer sat in a chair opposite the mayor in her distinguished-looking office. Neatly coifed and dressed in a gray-flannel suit, Mayor Feinstein said of her city, "It's one of the most diverse cities in the world today. There are many different people who live in this city, each one of whom marches to the tune of their own drummer."

That week, when the vote was taken, the mayor resoundingly defeated the recall attempt.

I have lived in San Francisco just three years, but I sincerely consider it to be one of the most beautiful and exciting places in the country to live. I just hope that my parents in Alabama didn't see your program, because if they did, the Cavalry will be arriving at my doorstep tomorrow!

Kelly McClay
San Francisco, California

Your representation of San Francisco was easily one of your shabbiest pieces of slanted sludge yet. It was foul, lurid, and outrageously distorted. You have denigrated one of America's great cities.

Max Goldstein
San Mateo, California

If you guys were in the house-cleaning business I would never hire you. You'd polish only the tops, sweep all the dirt under the rug, ignore all but the most obvious cobwebs, leave the glasses and dishware spotty, stuff the laundry in the drawers, and then spray liberally with lemon verbena scent before you left.

Barbara Turner
San Francisco, California

I'm a lifelong resident of Philadelphia and its environs, and love it. Still, this city has some liabilities I wouldn't wish on any city 3 thousand miles away.

Philadelphia's budget deficit is $200 million and rising daily; the commuter rails have been on strike for 6 weeks, and government officials are still yawning about how to end the walkout; crime may be going down a little, but the police presence—particularly in center city—is oppressive; speaking of center city, it's filthy—nothing like the clean streets of San Francisco; and there have been nearly a dozen school strikes in a decade. I won't even get into the gay issue; you certainly couldn't be a sheriff in Philadelphia and go on "60 Minutes" admitting you were gay.

Don't recall Dianne Feinstein—clone her! And send the copy to

Philadelphia City Hall. She's a lot better than our choices in this year's mayoral election!

Dawn Heefner Shlifer
Philadelphia, Pennsylvania

For Feinstein, against Feinstein. For guns, against guns. For gays, against gays. Communists, Libertarians, White Panthers, et al. They should name a granola cereal after San Francisco. What ain't flakes is fruits & nuts.

D. Shipley
Selbyville, Delaware

To all those people throughout the country who thought all the kooks and weirdos resided in Southern California, please check the geographical location of San Francisco. Thank you, "60 Minutes." And to all Northern Californians who advocate secession, you are free to go, providing San Francisco goes with you.

Everett F. Sanborn
Los Angeles, California

Although I've enjoyed living in Idaho this past year, I'm looking forward to moving back to San Francisco in a few weeks. Your story about "her honor," the recall, and the City summarized nicely for me many of the reasons why I anxiously await my move. Even the local news is interesting in San Francisco.

Marc Machbitz
San Francisco, California

If the "gay" element of San Francisco objects to seeing "Sister Boom Boom" representing a segment of their lifestyle on your coverage of "kook city," I don't blame them.

As a former active homosexual I condemn the segments of this lifestyle and its no-holds-barred drive for publicity. "Sister Boom Boom" is as patently offensive as a swastika spray-painted on a synagogue or a KKK cross burning!

You have shown the public the sick side of the lifestyle. I pray to God that it opens a few eyes!

Allan Benjamin
Los Angeles, California

I want to thank you for contributing to a national myth—that California is a cult of kooks, a haven for every weirdo and wacko this side of Greenwich Village. CBS really is a leader in this field of cute misrepresentation. For years one could turn on the CBS Evening News and see Harry Reasoner or Bernard Goldberg holding microphones up to hippies and homosexuals and health nuts and Walking Jukeboxes, signing off their segments picturesquely posed under the Golden Gate Bridge, shaking their heads (with affection, I must admit, as well as affectation) and smiling sarcastically into the sunset.

I want to thank those producers who, at some story-idea meeting, must have said, "Hey, we haven't scoffed at those flakes out in San Francisco for a while. Let's send Morley out there for some color, eh?!" It keeps our tourist business booming.

Michael Caleb Lester
Berkeley, California

The issues behind the proposed recall of San Francisco's mayor make Chicago mayoral politics seem comparatively mild. Here we only differ over such timeworn issues as race, snow removal, and potholes. The mayor was kind in labeling her city as "diverse." Chicago is diverse; San Francisco is a zoo.

David O. Moorshead
Chicago, Illinois

I believe that "60 Minutes," backed by a conservative, homophobic, and dangerously bourgeois view of the world, has violated a common journalistic standard of objectively examining all sides of an issue.

Throwing in Sister Boom Boom and a few other extreme types solely for better ratings and a few chuckles for "Middle America" is insulting, degrading, and serves only to promote greater prejudice and misunderstanding of other human beings. And normal gay people are certainly just as human, if not more humane, as the denizens of "Middle America."

Please, delete the titillating shock/amusement sidelines and stick to straight-forward journalism (and not, if you will, straight-forward prejudice). Remember: Prejudice promotes only violence and misery.

Peter George White
Beverly Hills, California

Where did you learn to count voting blocs? By your figures, of the 600,000+ residents of San Francisco and 400,000 registered voters, one-third of the residents and half of our registered voters would have to be gay. This is hardly the case. Your numbers are extremely high, even by gay political leaders' estimates. If you are going to glamorize the city's gays and enlarge their voting strength, first get your facts straight or your "straight facts" first.

Noah Griffin
Tiburon, California

Some day, investigative reporters will come to San Francisco and uncover the real scandal: There are straight people living here! The weirdest thing to discover about them is that, for the most part, they have a sense of humor and are open-minded enough to allow others in the city to clothe, coif, and comport themselves as they see fit.

Karrin Kain
San Francisco, California

As a former Texan, I remain convinced that the best place to be is Texas, where a "gay blade" is nothing more than a happy-go-lucky, sharp young man.

Bill Weaver
Middlebury, Connecticut

Let's extend that metaphoric title, ". . . Different Drummer." You guys spent all your time focusing on the snares—somewhat playful, more often irritating, with their constant complex cadences. Next time you come to San Francisco, give the entire band a chance. They're the ones who make the music.

Beth Seivert
Larkspur, California

Your portrait of San Francisco displayed the analytical depth of a tourist T-shirt. But that's OK. My son Joe has struggled 11 years to succeed in business printing T-shirts for such provincials as CBS editors, many of whom have come to roost here.

Al Ujcic
San Francisco, California

If you insist on interviewing only the looneys and crazies and oddities of the population of San Francisco, perhaps you should find out where they come from. Their real hometowns are, for the most part, back there in your part of the U.S. Maybe it's all that pollution you people have as a daily diet.

For you to deprecate this most beautiful City was most cruel and unjust. I would ask you to return and take the time to learn about the real San Francisco.

And another thing, Safer, didn't your mother ever tell you not to slouch in a chair and lounge like a lizard? Sit up and maybe you'll get the story "straight" next time. Shame on you!!

Doris M. Folk
San Francisco, California

You've Got Your Foot on My Neck
"Michael Doyle's Camden"

The story of the disintegration of Camden, New Jersey from a thriving industrial center to a decaying ghetto could be the story of many of America's older cities. The middle class, mostly white, has fled and the poor, mostly black, remain. With few resources and countless problems, such a city struggles mightily just to survive.

Harry Reasoner, in a segment aired March 20, 1983, found a spokesman to talk about Camden's problems in a personal and poignant way—Father Michael Doyle, the Irish-born parish priest who presides over The Church of the Sacred Heart in Camden. His small congregation is mostly poor, white, and elderly, many of them Irish. They are the ones who couldn't leave, or didn't want to leave, Camden. Father Doyle, who has a map-of-Ireland face and a brogue, claims he was assigned to Camden as a sort of punishment after anti–Vietnam War sentiments got him into trouble in another New Jersey parish. In Camden, he has continued to be outspoken, over a different "war"—the sociological battle between rich and poor.

Father Doyle was angry that Camden was being given a new state penitentiary. He was angry that Camden was getting the county sewage-treatment plant. ("Why isn't it in Tavistock? That's the biggest and fanciest country club we have in Camden County. Plenty of nice land there. They're just hitting little balls over it.") He was angry that people in the suburbs—some of them former Camden residents—were going to send their rubbish into Camden to be burned because they were out of landfill space.

People, he said, came back to Sacred Heart and said to him, "What's the smell?" And Father Doyle told Reasoner, "I represent a white man's church, and that's a limitation indeed. . . . For Ash Wednesday I would like to get this dirt that they say is in Camden and put that on the foreheads of society and say, '*Think!* You know, think—why is this place the way it is?' "

After watching your segment on Camden, we'll gladly put up with Buffalo jokes. Camden looks like a disaster area. Buffalo is beautiful, and we don't have to ask: "What the hell is that smell?"!

Szczepaniak Family
Buffalo, New York

Father Doyle complained about the middle class leaving Camden. I left Camden 3 years ago because it was no longer a safe place to raise a family or walk the streets. Also, why didn't Father Doyle tell you that when the middle class sold their houses, those houses were inspected by both the City (for a fee) and the FHA? In other words, the homes were in top condition when they were sold—we are tired of getting the blame for the condition that Camden is in now. The people with pride in themselves and their homes had to leave—we had no other choice.

Tom Shapiro
Audubon, New Jersey

For a decline of Camden the finger should be pointed at the blacks. Thirty years ago they moved in and drove out the WASPs, businesses, Italians, and Jews (in that order). They turned a beautiful city into a ghetto. For the past ten years they have been doing the same thing to Willingboro, a Camden suburb.

I'm not a racist nor a bigot. If anything, I'm just the opposite: a pluralist. But I'm sick and tired of a people who will not take responsibility for their own destiny, who are "professional victims."

During the 1960's this country turned itself inside out with the civil rights movement—which it would have been more appropriate to call the black rights movement. During the 1970's and 80's we went through an even bigger trauma called "busing." What did we get for our trouble? We get a bleeding-heart priest telling us that *we* are the cause of Camden's problems. Come on.

Robert Wagner
Lubbock, Texas

I'm trying to run a literacy organization that trains people to assist functionally illiterate adults. Most of our trained tutors are not willing to come into the city to do their good deed, because they are fearful and would rather tutor in the suburbs. But most of the students needing our service are in Camden, not in the suburbs.

Caryl Mackin-Wagner
Coordinator
Lindenwold, New Jersey

I travel over forty miles a day to work in Camden. I've been doing this for twelve years, working for the visiting nurse association and the school system. It's not because I can't find work closer to home, because as a nurse I find work just about anywhere I choose.

The people of Camden are fine people. They are warm, loving people with ambition, but they need help. Help to make their city the beautiful place it could be. This is why I work in Camden, traveling many miles a day. I may not get as much pay as I would working closer to home, but I get a thank you for my services with the children's smiles. They are nice, loving children who run to open the door for me even though I am of a different color than they are. Yes, I am a white nurse in a school of almost all black children and faculty. I have found something here in Camden that I never found working in the suburbs. It's kindness, warmth, a friendliness. Why don't you tell your viewers about Camden's good side?

Joan Currie
Mt. Holly, New Jersey

I am a young, college-educated, Black man who lived in Philadelphia all of his life (26 years), teaching in public schools. In 1976, I was given the opportunity to work in the Camden Public School District in a special program for junior high and high school students with learning and/or discipline problems. I chose to accept the job and to move myself and my family to Camden, a city with many problems perhaps (urban decay, poverty, crime, etc.), but certainly no worse than Philadelphia. There was something here in Camden that was different—a general feeling that there was a future here, that working together, there could be a turnaround in the problems, that what some people once considered "dead" could be brought back and redefined, enlivened, and even made prosperous—my wife, children, and I were all enthusiastic.

Living here for six years has not dampened that enthusiasm in the least, nor did "60 Minutes" piece. I no longer work for the Board of Education. Now, I run a growing marine services business—with customers who come from all over the South Jersey area (Camden included) and Philadelphia and its suburbs. I am occasionally questioned by people outside of Camden about the safety of the boatyard, about the vandalism that must besiege this place. We live and work in what some (Father Doyle perhaps) might consider an area once thriving and now "dead." The business does not suffer vandalism and never has. However, we receive customers here who are outraged at what has happened to their boats in other yards in the very best areas. They choose to come here—to Camden—to be safe, and to have their boats maintained.

Give us a fair shake—give us, the minority businessmen, the Whites who have come here to buy homes, the customers who have found sound businesses with which to deal—a chance to respond to the picture you and Father Doyle painted of us. It was a cruel blow to all that we have worked for and believe in.

Rodney S. Sadler, Sr.
Camden, New Jersey

Last week, there was a sale of city-owned properties in Camden. The conditions of sale required owner-occupancy within one year. A standing-room-only crowd of eager buyers grabbed every property available. Now they're preparing to move here.

They must know something you don't.

Lois M. Teer
Camden, New Jersey

I am so disgusted with your reporting-distorting techniques I can't bring myself to address you as "Dear 60 Minutes." There can be nothing dear about such destructive reporting. I am a fairly new resident to Camden and was willing to work and become involved in its rebirth. Fortunately for Camden I have enough sense to realize that there are two sides to most stories, and will continue to see the trees, despite what you tried to do to the forest. And in the future, when Camden doesn't have room for the sewage because of neighborhood encroachment, I know where I'll vote to send the overflow.

Jane Jones
Camden, New Jersey

Congratulations to you and Father Michael Doyle for an eloquent and scathing indictment of corruption and decay in Camden. It was magnificent.

In fact, the whole "60 Minutes" show was outstanding reporting. I hope it will be repeated.

Unfortunately, "60 Minutes" was preempted for over an hour here in the Camden-Philadelphia area by WCAU (Channel 10) due to the presentation of an unannounced college basketball game! As a result, I'm afraid many potential viewers were lost in the very area where your report mattered most. I think CBS should investigate why Channel 10 sabotaged the showing of such an important and timely show.

Richard O'Connell
Wyncote, Pennsylvania

If there were more priests as dedicated to the people and as real as Father Michael Doyle, I would still be attending the Catholic Church.

Florence Ordaz
Rockford, Illinois

For 13 years I've driven by the scenes of Camden that you showed last night, I know they exist. Years ago, you may have heard, Walt Whitman called Camden "the asshole of Philadelphia." He of sainted memory.

But I have an uncomfortable feeling of having been had. The scenes you showed were real, and while Camden does boast The Best Steak Sandwich in All of Existence (Donkey's Cafe, down the street from Lourdes Hospital—you *should* have mentioned it, if only for kickbacks from physicians: they're greasy as hell) it has little else to commend it.

EXCEPT. It is, for better or for worse, mostly worse, the home of a lot of people who care and hope and who have been sold down the river by a lot of other people. Corrupt politicians, inept businessmen. You let Mike Doyle tell you that these people had no cause for hope.

I think you offered great Irish television on the Sunday after Saint Paddy's Day by a charming Irishman who presides over what you showed as an almost lily-white Irish parish. He is of course a more romantic spokesman than Camden's mayor, Randy Primas—there are no black leprechauns that I know of.

But your story calls all of your reporting into question, at least in my mind. I know the Camden story, and you did not tell all of it. Where the human spirit exists there is hope, and you told nothing of hope. Where people hurt and cry out, there are still some people who will listen.

For example—forgive me for being parochial—while Mike Doyle spews his doom, the Lutheran Church is starting New Life Lutheran Church. Hey, we don't start new missions only in Suburbia, some of them are in Ghettoland, around the corner from West Jersey Hospital which you showed on Your Very Own Tee-vee. And though I've never meet Mayor Primas, my understanding is that he's the first honest mayor Camden has had in many years. He's got an incredibly difficult job (I think he cares!) and his victories aren't many yet.

But I attended, for example, the opening of a new building by a parishioner of mine—the best commercial photographer in south Jersey. He looked at the suburbs but decided against leaving Camden. In so doing, he showed a kind of commitment and guts that you totally decided to ignore.

Mr. Reasoner, you let yourself be sold a bill of goods by a leprechaun who decided to be charming to you. He told you a part of the story, and that part is true. But it isn't the whole part. I think Camden got a Black Eye from the Eye from Black Rock. You weren't fair, you weren't complete, but, more importantly, you weren't honest.

Rev. J. Bert Carlson
Mt. Laurel, New Jersey

Zurich

"What's Wrong with Switzerland?"

Harry Reasoner went to Zurich, Switzerland, to check out the rumor that this is one of the most boring countries in the world. Oh yes, he noted, it's true that practically every Swiss citizen who wants one has a job, there's hardly any crime, and the scenery is great.

But people don't seem very happy there. The head of the Swiss National Bank told Reasoner, "The typical Swiss is hardworking, a serious man, some people think much too serious. They don't laugh as much as other people." And sure enough, "60 Minutes" cameras recorded many dour faces in the streets of Zurich.

Order is important to the Swiss, said a Swiss journalist. They feel, he says, that "it's better to respect the order and to be unhappy [than to be] happy without respecting order." There were scenes of one of Zurich's clean, graffiti-free, beautifully designed trams. As the doors opened, the passengers filed out in neat, orderly fashion. Reasoner visited the two enormous underground kitchens civil-defense planners have built and stocked in case of nuclear war. A schoolgirl was shown being coached by a traffic guard in the proper, careful way to cross the street.

But signs of disorder were there. Young people have begun to protest. "The protestors adopted as their symbol the letter 'A' for anarchy," said Reasoner, "and call themselves 'the movement of the disenchanted.' " A militant fringe group called "The Chaotics" staged a "terrorist attack" on the "60 Minutes" camera crew, taking away their cameras, covering them with paint, and tossing bottles of Heinz ketchup. The attack lasted 10 minutes, and was captured on film by a freelance cameraman who sold his footage to "60 Minutes."

"We had come to Switzerland planning to do an essay on life in a boring country," said Reasoner. "We left thinking that if the alternative to boredom is chaos, maybe a little boredom isn't so bad after all."

Your understanding of Switzerland is epitomized by your charming notion that Zurich is its capital. The last man who espoused such an opinion was eaten alive by a big bad Bern.

Raymond and Heidi Johnson
Kenosha, Wisconsin

Check again on the flag of Switzerland. It is certainly not the flag you pictured beneath the title. That is the Danish flag.

Ann Gardner
Calgary, Alberta

In discussing the extensive civil-defense efforts of the Swiss, you observed that cans of food stored in underground shelters would be penetrated by radiation. You unfortunately are under a common misconception. Exposing sealed cans of food to massive doses of radiation will not contaminate them—only mixing the radioactive particles in with the food would do that. Massive radiation exposure can actually be used to preserve food. An all-out nuclear war may or may not be survivable, but the Swiss, who spend 100 times more than the U.S. on civil defense (per capita), are not fools.

Robert Ehrlich
Fairfax, Virginia

Having lived in Zurich for 2 years, I can verify that the Swiss are a strange, intolerable, and bigoted people. God help us if they should be the only survivors of a world nuclear holocaust. If that ever comes to pass, I pray I am 50,000 miles away from Switzerland.

Elvira Vigil-Ogard
Santa Fe, New Mexico

Your recent segment on Switzerland reinforced the suspicions of the wife of this American-born Swiss in re my dour disposition. Couldn't you have balanced the broadcast with a pastoral mountain scene?

Herbert G. Walther, Pastor
Trinity Lutheran Church
Danville, Illinois

Switzerland shares borders, languages, and cultures with several diverse neighbors. Zurich, the city you visited, lies in the teutonic north—only 15 miles from West Germany. Small wonder then that those "Swiss" traits you found included an absence of humor, an anxious obsession with order, and an open contempt for visiting Americans (in this case, CBS newsmen).

John S. Workman, Jr.
New Haven, Connecticut

When Harry Reasoner ended his segment on the tranquility of Swiss life by saying, "Chaos is better than boredom," you can bet he's never been a mother!

Shauna P. Faulhaber
South Bend, Indiana

Watching your presentation on Switzerland last Sunday, I felt sort of sorry for you because it seemed that you couldn't quite get off the ground. Apparently nobody had ever told you how it all began. So allow me:

When God created the Swiss he found that he felt particularly warm and loving toward them. (It seems we have to grant foibles even to the Lord . . .) And so he decided to be particularly good to them and said, "My dear Swiss, what would you like to have?"

And the Swiss replied, "Give us the most beautiful mountains in the world." And the Lord gave them the most beautiful mountains in the world. And then the Lord asked, "What else would you like to have, my dear Swiss people?"

And the Swiss replied, "Give us the most beautiful lakes in the world." So the Lord gave them the most beautiful lakes in the world. And then he asked them, "And what else would you like to have, my dear Swiss?"

And the Swiss replied, "Give us the most magnificent cattle of the world." And the Lord gave them the most magnificent cattle in the world.

But after all this creating the Lord felt a little tired, and he asked the Swiss to give him a glass of milk. They gave him a glass of milk from their wonderful cattle, and the Lord rested for a while.

And when he felt rested, he asked, "Is there something else you would want, my dear Swiss?"

And the Swiss replied, "Yes, Lord. Ninety-five cents for the glass of milk . . ."

So there. Now you know.

Gertrud M. Kurth, PhD
New York, New York

After having just returned from one year as a visiting professor at the University of Bern, Switzerland, we want to congratulate you and share a few more peculiarities of this beautiful country with you:

1. All new homes must be constructed with underground bomb shelters *in case of war.*
2. Women are not allowed to vote in certain cantons (states) and the right to vote is periodically denied to them by the voting men of that canton.
3. Do you want to know why the Swiss people do not smile much? There are only 3 television stations in Switzerland, all government owned: one in German, one in French, and one in Italian. Programs start about 5 P.M. and end about midnight. One day there was a special program that started at 6 A.M. It was a day in the life of a Swiss farmer. It had no narration, and consisted of following his activities—milking cows, eating meals, and plowing fields. It lasted until 5 P.M.
4. Automobiles are forbidden from having rust and must be inspected every 3 years.

It's a nice place to visit, but . . .

Bob and Nancy Kiel
Simsbury, Connecticut

How could you possibly do a program on Switzerland with the particular focus you gave it—serious mein, rules and regulations governing personal living, etc., and not mention the name of John Calvin!

Was your historian on vacation?

Jean Higgins
Smith College
Department of Religion
Northampton, Massachusetts

I was happy to see that you left-wing liberals (excuse me, Andy) had to admit that conservatism, at least, beats anarchy.

Since in all the years of your program I have never heard you admit that conservatism could beat anything, the Swiss segment was a refreshing change. Who knows, maybe someday you'll even admit that conservatism in America also has some good points. Maybe there is hope for you yet.

Jack C. Barr
Burlington, West Virginia

I am glad you thought Switzerland was conservative or boring. I wouldn't wish to disillusion a soul.

I lived there for about a year, teaching at an international boarding school. There were no gripes, and everything was spotless and in good repair. It was very relaxed, very stimulating, very cordial. When the kiddies graduated, they all got into the university of their choice.

I would sit looking out my window at the mountain across the valley sparkling with its glaciers in the sunlight like a diamond set in an emerald landscape, trailing waterfalls down its slopes. I could have stayed forever. I was never bored. Why didn't I write about my experience for everyone to know? Well, I followed my golden rule about all the little Shangri-las in the world I've discovered: I ain't tellin'!

Sara M. Drake
Lake Forest, Illinois

Polo, Illinois

"Small Town, U.S.A."

Americans have very mixed feelings about small-town life. We sentimentalize it in popular entertainment, applauding its preservation of old-fashioned virtues such as neighborliness and compassion. But there is also a thick vein of literature that depicts small-town residents as provincial, narrow-minded, and insular. Perhaps nobody has come much closer to the truth of the matter than novelist Doris Lessing, who once wrote, "The art of living in a small town is one of the most difficult to acquire."

Polo, Illinois, is a farming community of 2,600 about 120 miles west of Chicago. In 1978 Stuart and Trudy Stitzel moved there from Chicago with their four children. One of those children, who was 11 at the time "60 Minutes" told the Stitzels' story in October 1982, was severely brain-damaged. She weighed less than 30 pounds and required round-the-clock nursing care. Her parents had chosen to give it to her at home rather than put her in an institution. When they moved to Polo, reported Mike Wallace, they thought it would be "a good place for Susan to breathe fresh air . . . clear air, a clean start. But it hasn't worked out that way."

Stuart Stitzel worked as a teamster in nearby Dixon, earning $19,000 a year. In addition, Susan received $166 a month in Federal Supplemental Security Income because of her disability. But many of the Stitzels' expenses were not covered. In 1981, for example, they had to pay over $4,700 in medical expenses themselves.

Early in 1981 Stuart Stitzel asked the town of Polo for help in paying his water bill. He had occasionally received help from the town in previous years with heating bills, and sometimes even for food. The town knew a lot about the Stitzels. The state had asked for an analysis of their financial situation, and a professional budget planner for the University of Illinois had found that after basic expenses the Stitzels had $46.75 a week left over to feed and clothe their family of six. (The budget planner also said that by keeping Susan at home the Stitzels were saving the state nearly $20,000 a year.)

Two days before Christmas the town sent the Stitzels a notice telling them their water would be turned off if their water bill wasn't paid in full. On January 14, 1982, Stuart Stitzel made a $5 payment toward the outstanding bill of $672. Nevertheless, on the morning of January 15, in the middle of a blizzard, a water-department truck pulled up to the Stitzel house and disconnected their water supply. The temperature was below zero. Trudy Stitzel was unable to use the suction machine that kept Susan's lungs clear. It was 9:30 A.M. and Susan Stitzel hadn't yet had a drink of water or a bowl of the special cereal formula she ate.

The Stitzels were without water for one day before the City of Polo wrote up an agreement allowing them water and requiring them to pay $25 on their back bill. Stuart Stitzel signed it, even though he knew he couldn't pay it, because it was the only way he could get the water turned back on. Then, when he didn't keep up the payments, the town threatened to turn off the water again. Stuart Stitzel went to court and sued the town officials for $250,000. A federal judge ordered Stitzel to pay $5 a week, and restrained the city from turning off the water.

How could this have happened? Wallace's interviews with people in Polo were revealing. Two people on the street, their eyes narrowed, said, "If I didn't pay my water bill they'd turn it off," and, "They should have done it sooner, a lot sooner." The publisher of the local newspaper, which came out against the Stitzels though no representative of the paper had ever visited the family or seen Susan Stitzel, said, "It was brought to my attention through other modes of hearsay that possibly this family wasn't managing their money as well as they might have, in forms of trips to taverns and things like this." Stuart Stitzel denied it. "I never go into a tavern. I will have maybe a beer or two at night when the children are all put to bed. . . . And that is it."

The newspaper took a poll. Ninety-one percent of the

Polo citizens who responded said the city was right to turn off the water. Eighty percent said Polo would be better off if the Stitzels left town.

In late September 1982, a week before "60 Minutes" broadcast its report on the Stitzel family, Susan Stitzel died. Her funeral was held not in Polo but back in Chicago.

The Stitzels' lawsuit was settled out of court; Polo agreed to pay them $60,000 as a result of shutting off their water.

These later events were reported when "60 Minutes" rebroadcast the segment in July 1983. Eventually the Stitzels would leave town for good.

Over four thousand viewers wrote in about this segment. A number of them sent money to be forwarded to the Stitzels. The overwhelming majority of letters supported the Stitzels and denounced the town of Polo. However, in the selection that follows no attempt has been made to be statistically representative. Instead, pro and con letters are interspersed to illuminate the nature of the debate this segment sparked, a debate that goes to the heart of such issues as good versus evil, the roles of money and community in our lives, and human nature itself.

We are all in the midst of hard times, and I'm sure that patience is short with the problems of other people because we all have our own problems to deal with. But when people are in need and doing all they can do to better themselves, you don't help them by calling them deadbeats, or gossiping about them, risking the safety of their children, or suggesting they leave town. Where are the churches when these situations arise? From the way it looks, the town of Polo is so full of hypocrites that they don't need a church because they probably wouldn't go anyway.

Polo reminds me of a story written by the late Shirley Jackson, "The Lottery." In this story a whole town comes together to stone a woman to death so that they may survive. In this case the Stitzels' little girl was the town sacrifice.

Anthony M. Davis
Marrero, Louisiana

Those people were not "deadbeats," as the citizens of that small town would have you think. That was a family who cared enough about their little girl to take care of her themselves instead of institutionalizing her.

And what business is it of those townspeople's whether the family's bills are paid or not? Also, how did word get around town about this unpaid water bill? The father was working, making decent money, and the son also worked. It's not as if they were just sitting back, waiting for someone to give them handouts. I'm sure if they could have paid the water bill, they would have.

We've all heard about small-town people and how mean and nasty they can be to people who aren't one of "them." I really didn't think it was like that, but I was wrong. I wonder if those people saw themselves on TV. I hope they did.

Renee I. Martin
Danbury, Connecticut

We are former residents of the Polo area. I know the people you interviewed and they are not like you tried your very best to present them. People there do not care to be bilked by a deadbeat.

The idea that a family cannot make it on a $20,000 income plus hundreds and hundreds of dollars in aid besides is ridiculous. Their expenses for that child were not that great considering the fact her medicine etc. was paid for and they cared for her themselves. As far as I could make out her "special food" was cereal. Nothing else was mentioned. All her baths she needed per day were sponge baths. (I worked in a hospital as an aide and gave baths for reducing temperature and that is how it is done.) Not much water involved there. I was told distilled water was used in her suctioning machine. Don't know if that's true or not but they do not use a huge amount anyway.

As for the interview with the social worker we all know what a scandal the welfare program is. Any "facts" she presented were likely garbage. The "essentials" they figure a person or family needs are a far cry from what they really can get along with. A "bare bones" budget is called for—a roof over their heads and some plain food on the table (and no beers after dinner). We don't and can't afford such stuff. We are farmers and really live on "bare bones." We never go anywhere and do anything that costs money.

You liberals are all alike. You think everyone has to be given everything they desire. Going without what you can't afford is the way it should be and pay what you owe first.

Mrs. Fred Venhuizen
Muscoda, Wisconsin.

The United States is my adopted country, and I must say—without reservations—that Americans are the kindest and most generous (almost to a fault) people I have ever known. If the town of Polo has hardened toward this family, there must be A LOT more to it than was revealed in your report.

The funeral provided a clue: The Top-of-the-line coffin certainly gave me the impression that this family's priorities are NOT outstanding water bills.

Helga Spring
Eugene, Oregon

To not be sympathetic about the sad condition of the little girl would require a stone heart. But to allow a water bill to accumulate to over $400 with repeated warnings and appeals to pay something, was totally uncalled for. Cutting off service was proper and in the interest of all other taxpayers. When the father said he could not pay $5 per week on the bill, while he had "a couple of beers" each night to relax, proved they had no intention of meeting their obligations rationally.

To hear that the town had to pay that family of parasites $60,000 for turning off their water supply was nauseating. We wonder just what kind of a scam those deadbeats can pull on their new city. Certainly the old man can upgrade his beer to Wild Turkey until their ill-gotten windfall runs out.

Paul G. Jacobs
Austin, Texas

I must say I have never been so upset in my life! It is despicable the way the town treated this family. Mr. Stitzel had four children, one disabled, and earned $19,000 a year. That is nothing to raise a family on these days, especially when one of the children was so ill. It was alleged Mr. Stitzel frequented a tavern. Well, carrying a burden such as this, it's a wonder the parents didn't drink themselves into oblivion!

Mrs. David J. Pollack
Roswell, Georgia

I have lived in the town of Polo for twenty-six years. We raised our children here and there isn't a better town anywhere. I have given money to help a young girl have a liver transplant and another young child to have heart surgery. There is always someone ready to help someone else in this town.

The town of Polo is made up of many classes of people. Most of them are hard-working, church-going people. They are glad to help, but they don't appreciate leeches.

You were here in Polo about a week. You didn't see their home before that and you have no idea what they really are like. The Stitzels really cleaned their home up before you arrived. Fresh paint, new carpeting, garbage cleaned up. Do you know what the Stitzel kids are saying this morning? Quote—"As soon as the money starts rolling in we're moving to Florida."

You probably think I sound like a real hateful person. Well I'm not. I am gladly willing to help anyone anytime and I have. But it makes me boiling mad to be taken advantage of.

If you can afford to smoke, go to taverns (not in Polo—in Dixon) and pay for cable TV, you can show good effort by trying to pay your bills. I am not signing this as I do not want to be sued by the Stitzels. I'll sign it as

Mrs. X
Polo, Illinois

That little Illinois town: little town, little people, little hearts, little minds! That little-minded "Supervisor" speaking of the dollar which the Stitzels got together for charity: "Then they can't be too poor, can they?"

God! I'm glad the little town has to cough up $60,000! Little in every way!

A. E. Hinson
El Cajon, California

Polo, Illinois reminded me of a pop hit in the 1950s (sung by Gene Pitney). It was the title song from the movie *Town Without Pity.* It ended, "And it isn't very pretty what a town without pity can do."

Robert Sagan
Reno, Nevada

The reaction of the residents of Polo, Illinois to the plight of the Stitzel family was appalling; it was also an unparalleled example of small-town, opinionated bigotry. That the city leaders could not comprehend the staggering responsibility, let alone the emotional trauma, that is the result of caring for a totally disabled loved one leaves me with a feeling of indignation and sadness.

I will never again look upon my home state with the naivete of the young girl who believed it to be the best place in the world to live happily ever after. I hope the town of Polo will be more receptive to the next new family on the block.

Mrs. Sarah Laszlo
San Diego, California

As a Polo resident I am very upset and angry about "60 Minutes" portrayal of Polo people as uncaring. It just isn't true. They have been given loads of groceries from the Church of the Brethren and had *several* heat bills paid by St. Mary's Catholic Church, and had furniture donated to them. It was only the last heat bill that they were asked to pay. Their water should not have been turned off in January, but they *did know* about the arrangements before it got to the point where it was turned off. If water was that important to Susan's life surely that should have been something that they took care of immediately.

We realize the tremendous financial burden they are under, but it hasn't stopped them from having 3 cars, an Atari, cable television which costs at least $18 a month, and Calvin Klein jeans.

Joe and Betty Bleistein
Polo, Illinois

"60 Minutes" told the story I had hoped to never hear. It is the story of love as St. Paul told it to his friend Timothy (I Tim. 6:10): "For the love of money is the root of all evil: . . ." The raw drama of a community turning their backs upon human need of water is the greatest of evils, but a people so void of human kindness that they did not express a single word of loving concern for this family, is a sin against the Christ, and beyond the imagination of anyone serving the Suffering Servant.

Has Polo paid its water bill to the Lord? I hope they have, because they are going to need a friend, and they might find the well is dry.

Leon Peacock, Pastor
Chapel-in-the-Woods
Wimberley, Texas

I think the citizens of Polo should have their water checked, because there is obviously something in it that contaminates their sense of feeling for other human beings.

Kevin Dark
Blue Anchor, New Jersey

After listening to the townspeople and reading their facial language as you interviewed them, I realized there is a gold mine of work for me and my career in Polo. I'm moving immediately.

Marion J. Wolsey, M.Ed.
Counselor and Education
Specialist
Tucson, Arizona

Your criticism of the mayor, water department, and people of Polo, Ill. for disconnecting water service to the customer that had failed to pay their bill was totally unreasonable, unjustified, and displayed a lack of understanding or concern for the true responsibilities of government and its officials.

In the utility service field, we have to treat all customers uniformly with no special rates or rules for individuals due to race, religion, or economic status. If an individual needs help in paying for the services received, he or she should seek assistance from various social welfare agencies that are funded for this purpose.

James L. Hinchberger
County Sanitary Engineer
Butler County Water &
Sewer Dept.
Hamilton, Ohio

I was appalled! How this kind of cruelty can exist in America is beyond comprehension. Our town has over $70,000 in overdue water bills. No one's water has been shut off.

Cecile M. Begin
North Adams, Massachusetts

It is no coincidence that the narrow, mean, vindictive, uncharitable town of Polo, Ill. is near the little town that produced Ronald Reagan. They personify smug, selfish America at its worst. May God liberate us from these Neanderthals, soon. Our WORST are leading us astray! Christians they are NOT.

Mrs. Nancy Bey
Redwood City, California

First of all, sorry I don't have a typewriter. I make $22,000 a year, and I can't afford one. The people in Polo would probably say I don't know how to manage my finances.

I'm a home health-care nurse and I help people like the Stitzels keep their handicapped or seriously ill relatives at home. I wonder if other people understand what a financial, physical, and emotional strain this decision can place in a family and how wonderful it is at the same time, for the patient and the rest of the people involved.

Watching your program (and in my daily work) I realized no nursing home in the world will provide the individualized, loving nursing care that mother was giving to her child.

I believe the people of Polo, Illinois ought to be brought to trial for crimes against a human being. I'm glad I live in New Mexico. Authorities here are very cooperative when I call and inform them about utilities problems in my patients' homes.

G. R. Ludmer
Albuquerque, New Mexico

My hat goes off to the family of little Susan Stitzel for insisting on keeping her home where she belonged, and where she was loved, apparently very much.

I was moved to tears by this story. I think if Susan was ever aware of how little those idiots cared about her life, she probably died of a broken heart.

Sometimes I'm ashamed to say I'm part of the human race.

Mrs. Maria Walden
Milpitas, California

Polo's treatment of the Stitzel family reminds me of a Biblical scripture in 1st John 3:17 which reads, "But whoso hath this world's good, and seeth his brother have need, and shutteth up his bowels of compassion from him, how dwelleth the Love of God in him?"

Martha J. Adams
Portland, Oregon

I do believe that the water in Polo is a gift from God. I hope all those fine folks don't choke on it.

Kathee Anderson
Virginia Beach, Virginia

I am enraged at the inhumanity of the people of Polo. They appear to be robotic human shells, not unlike the unfeeling evildoers in the movie *Westworld.* The Stitzels were caught in a Kafkaesque nightmare.

As a psychologist, I feel I have a fair capacity for understanding human nature, but I'll never understand that kind of inhumane cruelty, prejudice, and lack of caring.

I wish for the Stitzel family love, abundance, happiness, and peace.

Barbara Schill
Malibu, California

All the people interviewed that had anything to do with cutting off the water at the home of Susan Stitzel—I hope they all contract syphillis and are allergic to penicillin.

Mary Catherine McGraw
St. Louis, Missouri

Well, I don't know where Polo is exactly, but you can bet on one thing, they better hope I don't happen to drive through by accident. I just might have to go to every faucet in town and turn it on in the middle of the night.

Guthrie Thomas
"American Folksinger"
Nashville, Tennessee

With the mills all down, our town of Klamath Falls opens their hearts and pocketbooks to those in need after only one newspaper article or a small spot on TV. Shame on you, Polo. Ill. And hooray for you, Klamath Falls.

Millie Colvin
Klamath Falls, Oregon

This Sunday evening, I was sitting at the kitchen table writing out my son's medical history for a college form, when suddenly my attention was drawn to the television. "60 Minutes" was doing a segment on the plight of the Stitzel family. I became so upset that I put down what I was doing to write this letter.

My heart goes out to this family, as I know what they went through when their daughter was alive. If you have never had a child who is handicapped or born with birth defects, as I did, you will never, never know the heartache and the devastating financial burden.

My husband and I incurred all our son's medical expenses ourselves. Looking back, I don't know how we did it but we did. There was no agency to turn to for financial help, no support group to talk with. Eighteen years ago my husband made only six thousand dollars. Our only saving graces were my son's doctor, who took only what hospitalization gave him, and he has done this for many years. The other was a kind and very generous pharmacist. This man, who was a total stranger to us when we first walked into his store, gave us medicine and medical supplies and any help he could at cost. He made no money on us. When we moved away from the area, he offered us financial help if it was needed. We will never forget him.

This September our son will enter college, and we are very proud parents. He will never be completely normal, but you would never know it to see him. Please for my son's sake do not use our name.

Name withheld by request
Ohio

"If any one gives you a cup of water to drink just because you belong to Christ, then I tell you solemnly, he will most certainly not lose his reward" (Mark 9:41).

The blessing of our God be upon you.

Michael Reardon, Actor
Currently performing in
"The Gospel According to Mark"
San Francisco, California

CHAPTER 20

Dear Andy

"Mail is an awful problem," he says sorrowfully, this man one viewer refers to as "the funny man who always complains about things at the end of the show."

But Andy Rooney means it. The thing that irritates him about the mail, and makes him feel guilty, too, is that he gets too much of it to answer. His audience from "60 Minutes" alone is vast; add to that the readers of his syndicated newspaper column and bestselling books and you have the potential for a monster load of mail.

He doesn't think he gets more mail than anybody at CBS News. ("I imagine Dan Rather gets the most, then Mike or Charlie Kuralt, Cronkite, and then maybe me.") But it's more than enough to "make me feel I should be doing something about it, even though I can't."

The trouble is, Rooney feels he should answer letters. Mike Wallace long ago gave up any notion, if he ever had one, that he should personally answer all his mail. And when Wallace does go to the trouble of making a personal reply, it's apt to be a very cordial but brief acknowledgment. Rooney, however, is a writer by trade. He regards a letter as a piece of writing, and knows that writing is difficult. And so he believes, "A good letter deserves an answer."

He has no secretary. Jane Bradford, his associate producer, does open his mail. She keeps a lot of it from him, sending it over to Audience Services, because she disapproves of Rooney's weird compulsion to waste time answering letters. But no matter what Jane does, every so often a fit of conscience comes over Rooney and he grabs a stack of letters, more or less at random, and answers them. He types the answers on his ancient Underwood, on the cheap yellow paper that used to be common in newspaper offices. The letters go directly from Rooney's typewriter to the lucky winners in this irrational lottery—no retyping, no letterhead.

Of course, he doesn't answer everybody; there aren't enough hours in the day. At first, he acquiesced to the CBS system and allowed Audience Services to write a form reply. "Dear Mr. Doe," it went, "Andy Rooney has asked me to thank you for your letter. He would, of course, prefer to reply personally but the heavy volume of mail he receives makes that impossible," etc.

He let them use that for a while, then it began to irritate him. He figured if people were going to get a form reply, it should at least be written by him. So he sat down at the Underwood and wrote the following, and instructed Audience Services to send it out, not retyped, just like this:

Memo to Letter Writers from Andy Rooney

There are good things and bad things about this recent well-knownness of mine. The money's good but there are problems. One of the problems is mail. I simply don't know what to do about it. I hate to think of all the people I've offended by not answering a letter they've sent me but I'm often getting as many as 100 letters a day. I hate answering letters anyway but even if I liked doing it, I couldn't answer 100 a day and do anything else.

This may be the most formless form letter you ever got but I'll tell you something I've thought about my mail for a couple of years now. We all make friends in different sections of our lives. We make them in grade school, high school and maybe college. We graduate, get married, take a job or move to another town and we make a whole new group of friends. We still like our old friends but our paths have diverged and we lose each other. We don't see our good old friends anymore. We make new friends and eventually part with them too. Our lives are compartmented and we have different friends in each compartment. No one can be friends all the time with all the friends he or she has made. Very often we lose track of them completely and can't even send them a Christmas card.

One of the best things about being in the public eye—and believe me, there aren't many good things about it—is that my old friends can find me and write me. Them I write back. Everyone else who writes me a good letter makes me feel terrible because I have to send them this.

Forgive me,
Andy Rooney

A little wordy, perhaps, and nakedly despairing, but still, a powerful and even intimate Rooney original. Mrs. Howard F. Donald, Sr. of Festus, Missouri, was one of those who received Rooney's "memo." She wrote back, "Enjoyed your letter, don't care if it came from a machine or whatever. I will continue to write, and you do not have to answer—OK?"

But Molly Courville of Santa Ana, California felt otherwise. She wrote to Audience Services, "I never wanted to be friends with Andy Rooney, and didn't much care to hear a diatribe on how they are found and subsequently lost, and found again. I suggest that if Andy Rooney finds being in the public eye so horrible he retire to some remote retreat where he need never be bothered by anyone again. Do I forgive him? NO!!!!!"

Perhaps you're beginning to see what it's like to be in the public eye. But then a letter like the following may come in the mail and make it all worthwhile. This gentle parody of Rooney's "formless form letter" was sent to him by Michael C. Chulay of Gendale, California:

Memo to Andy Rooney from a Letter Writer

There are good things and bad things about this relative obscurity of mine. The privacy is nice but there are problems. One of the problems is mail. Unlike you, I rarely get any. I've probably received as many letters in five years as you have on one of your 100-letter days. And I like mail, so much so that when I'm walking home from work the thought of a letter waiting for me is enough to put a spring in my step.

Of course, I realize that there's a world of difference between one letter from a friend and 100 from people you've never met. But if given the choice, I know I'd choose the 100 to none from a friend. Maybe that's worth remembering the next time you feel overwhelmed by their number—and the guilt from being unable to answer each one.

One of the best things about being out of the public eye—and believe me, the money is not one of them—is that I can write a silly letter like this and only one person reads it!

That's what *you* think, Michael Chulay.

As for the rest of his many fans, Rooney would probably be pleased if they continue to write and tell him how much they like him. But, please, if you don't get a personal reply, forgive him.

Fan Letters

In the introduction to this book, I promised readers they'd find no fan mail. I lied. Andy Rooney gets so many fan letters that a few of the best ones are worth sharing here, followed by a short selection of Mr. Rooney's letters from children. We begin with a fan letter from a new resident receiving instruction in this peculiarly American art form.

This is my very first fan letter. I'm a Hmong from Laos and a junior in high school. Our teacher wants us to write a fan letter, so I decided to write to you because every week you make me think about the things that surround me. I don't think you are a smart man but I think you are a funny man.

Ge Ly Boualong
Appleton, Wisconsin

The reason for this letter is that I failed to write P. G. Wodehouse, my favorite writer, before he died, nor C. S. Forester, my next favorite, before he died. So I'm working at telling people I enjoy their writing before they die and make it difficult.

I am an engineer and just write for fun (nobody will pay me for what I write unless it is an engineering report). I suspect that you write for fun and are continually amazed that you can ALSO get paid for it. I feel that way about engineering. It's like being paid for working out a complex and fascinating puzzle and coming up triumphantly with the right answers. Of course, not all engineers find answers, and some find the wrong answers, just as some writers are boring and some just plain wrong. These last are mostly Democrats like Arthur Schlesinger.

I like you on "60 Minutes." We change channels after "All Creatures Great and Small," which ends just in time to catch your segment. No need to tell the others.

Cary Hall
Hampton, Georgia

When you get older and become completely gray, I feel it is your duty to impersonate Santa Claus. Let's face it, your cheeks are like roses, your nose like a cherry. It's just what our children need in these troubled times: a Santa who is funny.

Stefanie Billings
Franklin, Maine

Some years past—1977, I believe—I was in my only blizzard. Normal activities were seriously curtailed for several weeks. (Arkansans are used to tornadoes. Some hours of severe terror, a couple of days of getting back to normal for the lucky ones.)

Anyway, to combat the boredom and cabin fever, I began rearranging my neighborhood to suit me. The neighborhood is filled with people who strike me as resourceful, intelligent, and entertaining, folks I wouldn't mind being weathered in with. Another requisite is that these people appear to be those who would find me reasonably interesting, and my dogs reasonably tolerable.

Sometimes when events seem to be approaching the unbearable, I have these folks over for croquet and gin-and-tonics. We crank ice cream under the mulberry tree if it's through bearing (otherwise, very messy) and talk about stars and space, books, words, dogs, music, ideas, and the human race. We talk about everything under the sun, except I'm not particularly enthused about sports.

The first person I moved into my neighborhood was James Cagney. He has lived here since 1977 and I've never considered moving him out. Others who have moved in and out have been Carl Sagan, Sister Elizabeth Kane, my stockbroker (he's in very great favor with me just now and still lives here), Margaret Thatcher, Mikhail Baryshnikov, Lily Tomlin, Millicent Fenwick, Barbara Woodhouse, P. G. Wodehouse, Kate Hepburn, Walter Cronkite, Dan Rather, and several others. I lean toward those people who could tell me things about the world for hours and hours.

I hope you don't find this too presumptuous, but I have just moved you into my neighborhood. Truth to tell, you're the first of my neighbors I've ever informed of this action. (If I could find an address for Cagney I'd write and tell him.) I'm giving serious thought to moving in James Galway and Randy Newman before the tornado season.

I hope your neighborhood has compatible, competent, intelligent, and interesting people in it, as mine does.

Tina Cook
Benton, Arkansas

Children (and One Exasperated Mother) Write to Mr. Rooney

The reason that I always watch your show is Andy Rooney. (Not that your articles are bad.) I think you should put him on every week. You might get more young viewers. I'm under ten.

Rob Sylvester
Naperville, Illinois

Dear Andy, How does President Reagan make a living? And does he have a 3 story house? Please answer these questions on: May 22, 1983.

Gina D. Tomonelli, age 10½
San Fernando, California

I liked your show about the keys because I know a lot of people like that, especially my teacher Mr. Sullivan. In the envelope I have sent you a ticket for Pizza Hut.

Jeff Coover
Shippensburg, Pennsylvania

I'm only 8½, but I'm on dishes this week. When I put the dishes away out of the dishwasher I have trouble getting the glasses all in the cupboard. We've got so many glasses it's hard to get them all in. Most of them are my little brother's. His name is Danny. While my dad was explaining how to put them all away I did an impersonation of you. I plugged my nose and said: Hello! It's Andy Rooney time! Tonight I'm going to tell you how to get all your glasses on the shelf.

I think it would be a good idea to do a show on how to put your glasses away. This might take a lot of your time but for my sake please do it. YOUR FAN,

Amy Victoria Johnson
Anaheim, California

Hi my name is Mary Nelson and I am 11 years old. One of your stories that I like best is the one where you said every winter you lose one glove out of every pair.

I think that at the beginning of 60 Minutes when everybody is saying their names, like "Hi my name is Ed Bradley," you should get to say "Hi, my name is Andy Rooney."

Mary Nelson
Cohasset, California

I'm 13 years old and I've been giving something a lot of thought. I *hate* Annie. I really mean it. She's always singing that song "Tomorrow" and saying that tomorrow something great will happen. Well, I tried to think her way but it just doesn't work. My dad told me that we are going to see a movie tomorrow. That was last week.

Elise Davis
Topanga, California

I am your Number 1 fan. I know you must receive a lot of fan mail that stakes the same claim. But I can prove it. You see, I'm an 8th grader at Crockett Junior High School in Odessa, Texas. I take Speech as an elective. We do a lot of skits. Naturally, I do Andy Rooney. I put talc in my hair and I have your voice down pat. I range from safety pins to pens. (I hope I have not taken any of your original subjects.) The class always applauds. I have competed in Speech contests where you had to write your own material and I have won half of them. I'm thinking of getting into your profession. I want to wish you a very happy new year. I'm very sorry of all my typos.

Jim Yeakel
Odessa, Texas

Did you ever wonder why some mothers long for asylum from their children who do impersonations of Andy Rooney? It wouldn't be bad if it were "a few minutes with Andy Rooney," but this 16-year-old windbag has been doing it for five days straight and even had the affrontery to call and leave a message in your voice on my telephone answering machine.

The solution would be to ignore this brain trust, but my efforts to stifle my laughter have been in vain.

I can't shut the TV off at 6:55 because he would see fit to fill in the vacant air time. So I now turn to the ventriloquist for this dummy and ask that you hammer shut his mouth till they bring back Pinkey Lee or the Mouseketeers.

Paula Sheffer-Fee
Sherman Oaks, California

CHAPTER 21

The Amateur Andy Rooney Hour

People like Andy Rooney a whole lot. True, the occasional viewer writes in to say something like, "Who is this guy and why is he telling us these dumb things?" but for every such sourpuss there are hundreds, probably thousands, who eagerly await his every Few Minutes.

Some of them like him so much they want to *be* him. They write him letters that start out, "Did you ever wonder about . . . ?" On occasion, these letters are actual scripts that could be read on the air—the right length, the right format. Some of them have been consciously written in hopes of a sale (Rooney does not buy material), but others, one suspects, are simply that old sincerest form of flattery—imitation.

Rooney reads some, but by no means all, of these letters. He says he almost never gets an idea from them. Most of them are forwarded to Audience Services for a form reply. Some of them are pretty bad. Some of them are pretty funny. He e's a selection from the latter category—a glimpse into what Americans wonder about, get irritated by, find frustrating, and get a laugh out of. (This chapter could have been called 'What Gets America's Goat," but then it wouldn't have had Andy Rooney's name in it.)

Fallout

How many times have you tiptoed out to get your Sunday paper and just when you feel you've made a perfect retrieval in your "jammies" all hell breaks loose—everything from small-sized TV program guides to Photo Film Processing Envelopes drop all over, causing the first and curse words of the day (a Sunday no less!).

It's even worse in the bathroom when you open Sports Illustrated, Penthouse, Redbook, Cosmo, or whatever, to have all kinds of small subscription inserts fall into your downed pants.

"Publication Fallout" is one of our REAL and PRESENT problems. It isn't something that COULD happen. It is causing more day-to-day frustration RIGHT NOW than all the possible tragedies any sign-carrying alarmist could forecast. Help us now, before we freeze our "jammies" off trying to pick up the crap that "Falls Out" of our Sunday papers and magazines every day of the year!

Bill Pettit
Willingboro, New Jersey

Have a Lousy Day

I learned a response to "Have a nice day." The next time someone hands you this line, with the usual dull tone of voice, try responding in a solemn tone, and with a stern look, "I have other plans."

I find it works very well, especially at K-Mart checkout counters.

James W. Harper
Des Moines, Iowa

Dew Drop By

In purchasing fertilizers for my lawn and garden I've observed that the only manure available at my local nursery is "steer" manure. I can't locate any bull, cow, heifer, or even cattle manure.

Questions:

1. Following the operation that a steer must undergo, do his droppings become superior to that of a cow or bull?

2. How do the commercial processors tell steer manure from bull or cow pies? Do they hire "watchers" to go into the pastures?
3. If they don't hire "watchers," what type of quality control is there? How am I sure that the product I buy is, indeed, dried, shredded steer pies?
4. How do they separate the pies? (I visualize an assembly line with a rubber-gloved person scornfully tossing cow patties off the conveyor belt.)
5. Finally, are there "waste" disposal sites for rejected pies? If there are, do they pose an environmental hazard to the public? If there aren't, where is it all?

I have a small cow herd and I always figured my steers brought more money on the market due to the quality of their meat. However, if it's really quality manure they're after, I may hire some "watchers" and bypass the middleman.

Ronald Smith
Bradley, California

Did You Say a Zillion?

I should like to ask you to mount a campaign that would greatly benefit the viewing public! Could you advocate changing either the word "million" or the word "billion," so that when we hear staggering statistics night after night which frequently are not enunciated too clearly, we would know just how much to stagger?

Carol Whitney
Batavia, Illinois

Lost Innocents

Is it not fascinating how everyone taken to court nowadays pleads not guilty? We must be living in a very pure world. Is it not fascinating how the lawyer for every person convicted promises the public he will appeal?

There was a wonderful picture in a recent issue of the *Boston*

Globe—5 innocents chained together being escorted to court, and every one with a leather coat over his head. In case of rain, I guess.

Margaret W. Tuttle
Concord, Massachusetts

A Trip to the Bank

I have just returned from A Trip to the Bank. It was the first trip *into* a bank I have made in years. (Most banking these days is done through your car window.)

It's incredible what banks can do these days. They have automated tellers, they can pay all your bills for you with one simple phone call, and they have more kinds of CD's and IRA's than you can shake a stick at. But have you tried to open a simple savings account lately?

Basically, I wanted to open a savings account to put away a little money for my children's tuition at a private school. I approached a cheery, bubbly young lady of about twenty and told her I wanted to open a savings account. She told me to have a seat, and then proceeded to explain that the bank had something called a Statement Savings Account, where you were given a passbook and each month you received a statement of your deposits and withdrawals. If your balance dropped under $200.00 there would be a service charge of $1.00 per month. I asked her if she had any other kinds of savings accounts and she said no.

I said, "Fine. I'd like to open an account, but I only have $111.00 in checks here." She said, "Then you can't open an account, because you have to have $200.00." And I said, "Do you mean to tell me that I have over $100.00 worth of checks in my hand, I want to open a savings account, and you won't take my money?"

She looked a little perplexed over that, as if it almost made sense even though it was contrary to bank policy, so she called over another officer. He said, "Yes, I have opened accounts for people with less than $200.00, as long as they have over $100.00 and they are advised that they will be charged $1.00 for the first month."

I said, "Fine. Let's open the account." She said, "Do you have an ID?"

I was prepared for this. Before I left the house I found my Social Security card. But the lady said to me, "We don't accept Social Security cards." So I said, "What do you accept?" And she said, "A driver's license." I searched for my driver's license, which I hadn't used for a while, and I couldn't find it, so I asked, "What else do you accept?" And she said, "Some sort of ID with a photo on it."

Well, for the past ten years I have been staying at home taking care of my children and being a homemaker, which I know is frowned upon these days. And I never needed an ID with a photo because my husband and kids always knew who I was. (Although I wasn't always so sure.) So I went back and searched some more and I finally came up with my driver's license. Maybe at this point I should have felt a little more secure knowing that my bank won't take money from just anybody.

She took my pittance with a frown of disdain and told me she would get a receipt from the teller. When she came back, I asked whatever had happened to the old-fashioned idea of bringing your kids into the bank with $5.00 and having them open savings accounts in their names so they could learn the value of saving. She looked at me with that puzzled expression again and said, "They don't do that anymore."

Maybe I know why "they don't do that anymore." A year or so ago I heard about a young boy who opened a savings account with $10 that had been given to him. He didn't touch the money for a year, and then he went back to make a withdrawal so he could buy his mother a birthday present. When he got to the bank he found that they had service-charged his entire account away, and he was left with a zero balance as a reward for his thriftiness.

I hope that this story might give you an idea that you can use on "60 Minutes." I even hope that you might pay me $1.00 for it so I can recoup my losses from opening the savings account. If you can't, I'll settle for an autographed picture.

Margaret Chasalow
Maplewood, New Jersey

Buttons

Why do men's and women's clothes button differently?
Which way do unisex clothes button?

Have any of the women's lib organizations taken a stand on these issues?

Hayes G. Garver II
Monterey Park, California

Running into Rooney

A lot has been written about how difficult a celebrity's life can be due to lack of privacy. But I've never read (even in the last column of the last page of the *Enquirer*) a word of sympathy about the hapless anonymous passerby and what we have to endure when confronted by a famous person.

The assumption has always been that we, the non-celebrities, can't wait to happen upon you, the celebrity. Let me put you and Rona Barrett straight about this.

Yesterday morning—early, very early—I, like so many other working commoners, was struggling to put one still-sleeping foot in front of the other in the direction of the newsstand. As I lumbered across 63rd St. I glanced into a car waiting at the light (I always glance into the cars at that light to see if everyone's as miserable as myself to be wrenched into reality at that ungodly hour. They usually are) and saw "Hey, that's—no it isn't, but wait, it *is*—what's-his-name—'Sixty Minutes'—Mickey Rooney. No, Andy Arledge—no, Andy Rooney."

Now, you shouldn't take this personally, but you were not foremost on my mind that rainy morning. In fact, you hadn't been foremost on my mind since maybe two Sundays ago. That's when I flipped you on for lack of a good old Sunday movie and muttered to my wife about What-ever-happened-to-good-old-Sunday-afternoon-movies?

What *was* foremost on my mind was the newspaper and, please God, some coffee. I didn't ask *you* to stop for that light. But you did anyway. You could have been some poor anonymous meatpacker from Queens on your way to a meatpacking plant. But you weren't. You were what's-his-name and I had to do something.

Now this is the point: Being surprised by a celebrity is like being surprised by your mother-in-law. You don't want to say anything but you have to. You can't ignore her. Her feelings will be hurt. She'll feel unloved. And you can't go home and tell your wife that you just ran into her mother in the middle of New York but didn't even say, "Hi, Mom."

Famous people, like mothers-in-law, are not chosen. They are plunked into our lives by some great power (CBS?) beyond our control. When you see one, you don't just keep walking, even if you want to. Think of the wife, the kids, the boss, the dinner guests. What will you tell them?

So there on 63rd St., in the early morning drizzle, I cranked up as much of a smile as I could, stopped in the middle of the intersection—almost being run over by the meatpacker from Queens—and croaked out, "Hello, Mister Rooney. . . . I really love your show!" (Just like I would tell my mother-in-law about her superb meatloaf. I mean, I had to think fast. It isn't *your* show—it's CBS's. It's not a show—it's a program and you have a segment. Sometimes I like it, sometimes I don't. I've never *loved* a show since Sid Caesar. But I had to think fast.)

You smiled. The light changed. And I ran to the curb for my life. I felt like an idiot.

I looked around. There was no one to tell. Besides, who cared? The meatpacker didn't care. I didn't really care either. I told my wife when I got home. She didn't care. I don't think you cared—and I almost got run over.

So maybe, someday, some sympathetic soul will write a few gentle words about how difficult it is being an anonymous guy in a sea of hungry, privacy-invading celebrities.

Peter H. Bemis
New York, New York

Computer Complex

I just keep pushing along, workday by workday, without causing waves, nor any great honors of one sort or another. That is OK with me, but why are the true and simple now made so complex?

For example, I used to know all the people in our offices. When I needed to check on things in another department I knew I could talk to Harry and he could tell me how it is done, where it is at, who has it, or who does it, and this information was good for years. Also,

Harry would always tell me other things. His tomatoes are prize this year because of the new fertilizer he used, or maybe about his Dahlias or about his weird uncle.

Now the computer has taken over and more and more unfamiliar faces turn up. The computer snowball has rolled Harry way back into a corner. None of the unfamiliar faces seem to know the answers to my questions. In fact, I feel like an alien when I pick my path between desk after desk. How did this thing get so big? Why do I get the same answer each time: "We don't do it that way any longer," "The computer won't take it," "It's not the right code," "The internal terminal department does that."

When I finally leave, very quietly too, and sit down at my old desk, in my wornout chair, I finally realize that I missed the latest news from Harry. How about his weird uncle? Are the slugs or root weevils worse this year in the garden?

Jerome Anderson
Seattle, Washington

Shoelace Peril

The other day I wandered into the neighborhood shoe repair store to replace a pair of common shoelaces.

What I purchased, as you can see, was a "not-so-common" pair.

Could you possibly reveal to all of us what Neohide means by "Impregnated Fabric"?

Until we find the answer to my query, I hesitate to give a good pull on my string.

Donald C. McGovern
Los Angeles, California

Oil Dilemma

I have just finished changing the oil in my car (I prefer to do it myself). And the question is, What can I do with the old oil and still be a law-abiding citizen?

The EPA says I can't bury it.

The NCFD (No. Cumberland Fire Dept.) says I can't burn it.

My MD says I can't drink it.

The SPCA says I can't feed it to my cat, and if I save it the IRS will slap a 10 percent tax on it.

So what the hell should I do with it?

I don't understand why all these people are coming into this country to be free.

J. Norman White
Cumberland, Rhode Island

Two Cents' Worth

I recently bought several 2-cent stamps to complete the postage for all those 18-cent stamps I bought but never used. Have you taken a good look at a 2-cent stamp? The "inspirational message" they've printed on it reads, "Freedom to Speak Out—A Root of Democracy."

Is this what's known as getting your two cents' worth? Is this the postal service's idea of a joke? If it is, you've got to notice that today you have to pay twenty cents just to get your two cents in.

C. W. Murry, Jr.
Kirkland, Washington

What Time Is It, Anyway?

It roils me, and I think it will you, when you note the trend towards designing the faces on expensive watches with no numerals whatever. Apparently the rich can disregard the importance of any particular moment. Or they enjoy the cerebral exercise of calculating an estimate.

Take this advertisement with the headlines: "The World's Ultimate Sportwatch: the Piaget Polo Day/Date." It shows the owner of one dashing down the field. He needs to know how many minutes and seconds of play are left in the scrum or round or whatever time period they use. He would have to pull his pony to a halt, dismount and put on his glasses, and then try to arrive at a general estimate.

Now look at the Concord. It should be the source of even greater frustration. The dial—if you can call it a dial—not only lacks numbers for the hours but also the gradation mark for minutes and a second hand. They say in the copy: "It makes the bulky sportswatch passé." It should be just the thing for the official timekeeper at the

Olympic sprints. Right? He might do about as well if he used a sextant.

Bob Herz
Dallas, Texas

A Good Question

What I would like to know is: To whom does the U.S. owe the national debt and to whom is the interest paid? You do lots of research so will know.

Mildred Tessmer
Janesville, Wisconsin

Of Envelopes and Planes

Firms will extend you credit, thus inferring that you have brains enough to earn the money to pay them back. Why then do they hint that you are an imbecile by sending with their bill a return envelope on which a small rectangle advises, "Place stamp here"? As if, for all your financial responsibility, the datum that postage should go on the envelope's upper right corner is too great a burden for you to hold steadily in mind.

Some go further. They flesh out these instructions with the observation, "The Post Office will not deliver mail without a stamp." True enough. (They might also say that the Post Office might fail to deliver it if it were *covered* with stamps.)

Another irritant has to do with air travel. For a while there, I believed that passing decades had degraded the romance of flight to the equivalent of airborne bus travel. Actually, it is worse. As far as I know, when the need arises, a bus will stop somewhere so passengers can get out and have a meal of sorts. They aren't constrained on board to eat without moving above their wrists. Bus lines do not needle you with a torrent of mindless noise. There is no Muzak. The driver does not ramble on jovially, telling you how far he is off the ground or what you might see out the window if you were there and wanted to look. And they don't have a stewardess who, through an amplifying system badly in need of servicing, passes on to you a paralyzing babble of instructions.

These injunctions are always drawn out to senseless length. Simple meanings are never matched with simple sentences; where four words will do, ten are used. Whoever invented this awful prose surely knew that it would be repeated ad nauseum on every flight. One suspects a bureaucrat, a teacher of education, or maybe an unfrocked tax lawyer—someone professionally blind to the risk of being a pain in the ass. As for the spiel on the use of oxygen masks: Since one may presume a disaster would be under way, surely the advice to breathe normally is overoptimistic.

Morrison Stiles
Charlotte, North Carolina

Chicken de Fright

I have always been a giblet lover and during my forty-plus years of buying and preparing fresh, whole chickens can't help but be amazed by the many ways they are divided and placed in those little bags.

Just to give you an idea, if you are lucky you might get one-half neck, one-half gizzard, one-half liver or three gizzard pieces, one-half heart or one neck and one-half liver, one-fourth gizzard or something you don't even recognize and on and on.

Today, I hit the jackpot. One neck, one-half liver, one-half gizzard, and ten hearts. Wouldn't medical science love to have known about that chicken?

Ruth M. Whiteside
Warren, Indiana

Burny Beans

It used to be that if you ate an orange jelly bean, it would taste like orange. A black jelly bean would be licorice, a yellow would be lemon, etc. But they have changed jelly beans. Now, white jelly beans are frequently licorice, or maybe yellow jelly beans are licorice. The red ones used to be cinnamon and really burny. But I haven't seen many of them lately. If you ask the people responsible for jelly beans, you never get a straight answer. Sometimes they add spices and herbs to jelly beans. This changes their true flavors, but they say they are more nutritious. They should be sold only by health food stores.

I'll bet you if Ronald Reagan ate a black jelly bean and it tasted like vanilla, things would really change.

Edgar Cornelius
Pittsburgh, Pennsylvania

How I Am

When you are as old as I am you'll be aware of the misuse of the question, HOW ARE YOU?

It's not used as a question any more, it's merely a greeting. In a few weeks I'll be 87, and when I'm greeted so 5 and 6 times a day, I hate it!

I don't want to have to think "How I am." Also, if I really tried to tell the greeter how I am, physically or mentally, the inquirer would be bored or horrified.

Yesterday I counted the number of times I had been so greeted and by noon there were 7.

I have various replies. I start with a courteous "Fine, thank you." Next I say, "Oh, all right, I guess." Then I retreat to "Well, haven't thought about it." A little irritation causes me to reply, "I don't know." Until finally I simply say, "Don't ask me."

Am I unreasonable? When you reach my age try to answer my question. I may still be around.

Margaretta Wegner
Stockton, California

Socket to 'Em

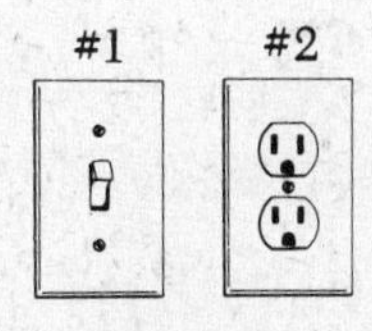

When a new house is constructed, the electrician installs item #1 about four feet above the floor next to the bedroom door. Then he invariably proceeds to install two or three of item # 2. Before he installs #2 however, he determines where the head of the bed will be and he places it exactly in the center of the headboard about 6″ or 8″ from the floor. Next he visualizes where the dresser will be placed, and places it in the middle of where that item goes 6″ or 8″ above the floor. He installs the last of #2 in the room behind where another wide piece of furniture will finally be placed.

Just thought this was worth passing on to someone who might understand.

J. Kyle Senter
Elizabethton, Tennessee

Is My Pillow Getting Smaller Or is My Head Getting Bigger?

I am a retired buyer and dept. manager for home furnishings. I am enclosing an old newspaper showing prices in former years. Over the last 40 years I have seen so many shortcuts being taken in the making of many items; I will list only a few.

- Men's suits—shorter pants pockets
  —narrower belt loops
  —no coin pocket in coats
- Men's shirts—buttons only to the waist—previously two buttons below waist
- Bed pillows—made 1″ narrower and 1″ shorter than original
- Jute instead of burlap used in carpet and linoleum
- The length of cords on table and floor lamps shortened about 18″
- Drapery and upholstery fabrics changed from 48″ width to 45″
- Webbing on aluminum outdoor chairs from 6 to 8 strips to 5 strips

We must get back to the days of producing and giving real value in products.

David P. Snyder, Sr.
Allentown, Pennsylvania

The Soft Life

I became a pilot because as a kid I saw that pilots didn't seem to do much all day except lean against their airplanes talking to the girls or sleeping under the bottom wing. Occasionally they'd hop a ride or do some aerobatics to impress the girls. They also wore boots, leather jackets, helmets, scarves, and wings. They seemed to get a lot of money for the little flying they did. Best of all, they never had to work, so I became a pilot and it was pretty much like I thought it would be when I was a kid.

Through all these years, I smugly went about flying airplanes and impressing the girls and feeling sorry for less fortunate individuals who had to work for a living on the ground.

I was secure psychologically until I saw you on "60 Minutes." Then the realization hit me hard and was unsettling. Here I saw a gray-haired guy who had it easier, obviously made more money, and seemed to be having more fun than me.

What you do looks so easy—saying and writing funny things—I'd like to get a job like you, doing the same and making some easy big money. Please advise how I can do it. Of course, I'll keep on flying because us pilots get all the girls.

Thanks for your help, Mr. Rooney. If you want to call (don't call collect) ask for Homer, my copilot. He'll come out to the airplane and wake me up.

Capt. Ernie Stadvec
c/o The Airport Bar
Akron, Ohio

Notemaker

I write a lot of notes to myself. My memory is not as good as it used to be, and there are a lot of important things I don't want to forget. Here is one I wrote to myself: It says, "Screwdriver." I can't remember why I wrote it.

There is a note I wrote pinned to the wall of our guest room. It says, "Change wallpaper and paint woodwork." It's there to remind our guests that the next time they come for a visit the room will look a little better. The paper is starting to turn yellow. Some of our guests have seen it three or four times.

Most of my notes are ideas about things I might want to write about or talk about someday. Here's one that says, "How do washcloths get dirty?" Here is another one: "If running water will wear away the mightiest mountain, why is the sink in my bathroom always plugged up?" Or, "Why does the wind blow through our house in the wintertime when all the windows are closed but not in the summertime when they are open?"

All the notes I write to myself are important, although I will admit some of them are more important than others. Here is one that says, "Call Mr. Jones at the IRS." It is written on the back of a postcard that is postmarked 1958. I guess that must have been one of the less important ones.

Gilbert Barnhill
Minneapolis, Minnesota

Mike and Morley

I have just finished watching "60 Minutes" and I would like to ask you a few things. Tell me—Why do you wear your watch with the face on the inside of your wrist? Do you like to twist your wrist every time that you look at your watch? Tell me—Why is Morley Safer such a poor dresser? I hear that he wears Savile Row suits, but to me it looks like he purchases his wardrobe at Sears. Some of the shirt/tie combinations he comes up with are unbelievable. Tell me—When will Mike Wallace learn to tie a proper knot on his necktie? I have never come across another man who wears a 5-inch knot in his necktie.

Tom Dugan
Watchung, New Jersey

Don't Ask

Just as a matter of curiosity, my mother, who is a perfectionist when it comes to German cooking, would like to know what you meant when you said that ". . . the Germans cook about as well as the British," in your segment about cookbooks.

Your reply would be greatly appreciated, since we are both devoted fans of yours.

Eleanore Baal
Larchmont, New York

Junk Mail

I drafted the attached letter. It goes to every piece of junk mail I get that doesn't interest me and has a postage-paid return envelope.

Dear Solicitor,

I am returning your postage-paid envelope without a response because I do not need the product or service which is the subject of your solicitation.

I could have simply deposited your literature in the trash can but that would only add to the gigantic refuse-disposal problem we are experiencing in my area. It seems to me the wiser decision is to let the creative minds of your company have the opportunity of deciding what to do with the refuse. It's clear to me the high quality of imagination that went into the development of your letter and brochures belongs to minds of far superior intelligence than I, for whom this kind of problem solving should come easy.

Secondly, I feel it is my patriotic duty to help my government whenever I can. As you know, the U.S. Postal Service has had some lean times of late. By returning your envelope you will have paid postage both ways thus increasing revenues for the U.S. of A. I call it the trickle-up theory although I do not expect to receive any commendations from Ronnie and the gang. But that's OK. I feel good about it and that's all that matters. I sincerely hope your company is just as patriotic as the undersigned.

Art Rodich
San Jose, California

All Bests to You

I enjoyed your dissertation on the toothbrush Sunday night. I think your idea of calling it a teethbrush is completely rational. That's what I call mine.

I keep my teethbrush in the baths room in a drawer with my feet powder, nails file, and extra hands towel. But when I get a new teethbrush, I put the old one in my tools box. I use it for cleaning sparks plugs and like to keep it handy with my lugs wrench and tires gauge.

I'm looking forward to your next program. You're a great gags writer.

Kirkham P. Ford
Paris, Tennessee